The **Politically Incorrect Guide**™ to
Darwinism and Intelligent Design

The Politically Incorrect Guide™ to
Darwinism and Intelligent Design

Jonathan Wells, Ph.D.

Since 1947
REGNERY
PUBLISHING, INC.
An Eagle Publishing Company • Washington, DC

Cataloging-in-Publication data on file with the Library of Congress

ISBN 1-59698-013-3

Published in the United States by
Regnery Publishing, Inc.
One Massachusetts Avenue, NW
Washington, DC 20001
www.regnery.com

Distributed to the trade by
National Book Network
Lanham, MD 20706

Manufactured in the United States of America

10 9 8 7 6 5 4 3 2 1

Books are available in quantity for promotional or premium use. Write to Director of Special Sales, Regnery Publishing, Inc., One Massachusetts Avenue NW, Washington, DC 20001, for information on discounts and terms or call (202) 216-0600.

For Phillip and Kathie Johnson

CONTENTS

❖❖❖❖❖❖❖❖❖❖❖❖❖❖

Chapter One

WARS AND RUMORS

"Evolution Wars" declares an August 2005 cover of *Time* magazine. In a parody of the Sistine Chapel, the bearded figure of God points down at a chimpanzee contemplating the subtitle of the cover story: "The push to teach 'intelligent design' raises a question: does God have a place in science class?"[1]

In March 2006, the American Association for the Advancement of Science issued an urgent "call to arms for American scientists, meant to recruit troops for the escalating war against creationism and its spinoff doctrine, intelligent design."[2]

Controversy over Darwinian evolution has been simmering for decades, and now it has erupted into a full-blown culture war between Darwinism and intelligent design. Pennsylvania State University anthropologist Pat Shipman calls intelligent design "horribly frightening" and writes: "I know that I and my colleagues in science are being stalked with careful and deadly deliberation. I fear my days are numbered." According to Marshall Berman, past president of the New Mexico Academy of Science, intelligent design "threatens all of science and society." Brown University Darwinist Kenneth R. Miller says, "What is at stake is, literally, everything."[3]

This sounds like more than a war of words—and it is. But it turns on the meanings of some key words, so let's begin by looking at them.

Guess what?

- The controversy is not over evolution—which can mean simply "change over time"—but Darwinism, which claims that design in living things is just an illusion.
- Intelligent design is not biblical creationism, but a scientific theory based on evidence from nature and consistent with everyday logic.
- Some Darwinists pretend they're just selling students on change over time when they're really peddling much more.

1

Evolution

"Evolution" has many meanings. In its most general sense it simply means "change over time." The present is different from the past. No sane person rejects evolution in this sense.

Refining the meaning slightly, anthropologist Eugenie C. Scott writes: "What unites astronomical, geological, and biological evolution is the concept of change through time. But . . . not all change is evolution, so we must distinguish evolution as being cumulative change through time."[4]

Nobody rejects evolution in this sense, either. Our grandparents had a perfectly good word for it: they called it "history."

In biology, evolution takes on additional meanings. Some biologists define it as "a change in gene frequencies over generations." Like "change over time" or "cumulative change over time," evolution in this sense is uncontroversial. My genes are different from my parents', and my children's genes are different from mine. So what?[5]

Charles Darwin's term for biological evolution was "descent with modification." When used in a limited sense, however, even this is uncontroversial. Like a change in gene frequencies, descent with modification happens every time a child is born. Breeders have been using artificial selection to produce descent with modification for centuries—within existing species. Natural selection has also been observed to do the same in the wild—but again, only within existing species.

So nobody in *any* field quarrels with "change over time" or "cumulative change over time." And nobody in biology doubts

Darwin + ism =

Darwinism consists of the following claims: (1) all living things are modified descendants of a common ancestor; (2) the principal mechanism of modification has been natural selection acting on undirected variations that originate in DNA mutations; and (3) unguided processes are sufficient to explain all features of living things—so whatever may *appear* to be design is just an illusion.

"change in gene frequencies" or "descent with modification" within existing species. Even hypotheses that some closely related species (such as finches on the Galápagos Islands) are descended with modification from a common ancestor are not particularly controversial; they generate more debate among evolutionary biologists than they do among biblical creationists, since Genesis states only that God created certain "kinds."

But Charles Darwin claimed far more than any of these things. In *The Origin of Species* he set out to explain the origin of not just one or a few species, but *all* species after the first—in short, all the diversity of life on Earth. The correct word for this is not evolution, but Darwinism.

Darwinism

Darwin wrote in *The Origin of Species*: "I view all beings not as special creations, but as the lineal descendants of some few beings" that lived in the distant past. Darwin believed that living things have been modified primarily by natural selection acting on random variations—survival of the fittest. "I am convinced," he wrote, "that Natural Selection has been the most important, but not the exclusive, means of modification."[6]

According to a 1998 booklet published by the U.S. National Academy of Sciences: "Organisms in nature typically produce more offspring than can survive and reproduce given the constraints of food, space, and other resources in the environment. These offspring often differ from one another in ways that are heritable—that is, they can pass on the differences genetically to their own offspring. If competing offspring have traits that are advantageous in a given environment, they will survive and pass on those traits. As differences continue to accumulate over generations,

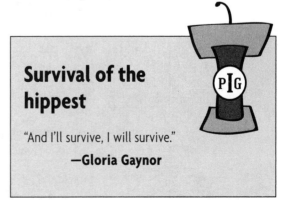

Survival of the hippest

"And I'll survive, I will survive."
—**Gloria Gaynor**

populations of organisms diverge from their ancestors. This straightforward process ... has led the earliest organisms on earth to diversify into all of the plants, animals, and microorganisms that exist today."[7]

Although the origin of life is often included in discussions of evolution, Darwin's theory applies only to living things. Darwin speculated that life may have started in "some warm little pond," but beyond that he had little to say on the subject. It seems likely that the first cells were bacteria, but as Harvard biologist Marc W. Kirschner and Berkeley biologist John C. Gerhart wrote in 2005: "Everything about evolution before the bacteria-like life forms is sheer conjecture," because "evidence is completely lacking about what preceded this early cellular ancestor." In any case, Darwinism does not include the origin of life.[8]

Nineteenth-century Harvard botanist Asa Gray argued that biological evolution was guided by God. Gray advised Darwin to assume "that variation has been led along certain beneficial lines. Streams flowing over a sloping plain by gravitation (here the counterpart of natural selection) may have worn their actual channels as they flowed; yet their particular courses may have been assigned."[9]

Darwin wrote to Gray that he was "charmed" with the stream metaphor, but he concluded his next book, *The Variation of Animals and Plants*

There isn't a Church Lady in intelligent design

Intelligent design (ID) maintains that it is possible to infer from empirical evidence that some features of the natural world are best explained by an intelligent cause rather than unguided natural processes. Since ID relies on evidence rather than on Scripture or religious doctrines, it is not creationism or a form of religion.

Under Domestication, with an explicit rejection of Gray's view. Using the metaphor of a house built with rocks found at the base of a cliff, Darwin explained: "The fragments of stone, though indispensable to the architect, bear to the edifice built by him the same relation which the fluctuating variations of each organic being bear to the varied and admirable structures ultimately acquired by its modified descendants." Thus "in regard to the use to which the fragments may be put, their shape may be strictly said to be accidental."[10]

In Darwin's metaphor, of course, the architect is natural selection, though he insisted that "natural selection means only the preservation of variations which independently arise." Darwin concluded: "There seems to be no more design in the variability of organic beings, and in the action of natural selection, than in the course which the winds blow." Although "I cannot look at the universe as the result of blind chance," he wrote, "yet I can see no evidence of beneficent design, or indeed of design of any kind, in the details." He was "inclined to look at everything as resulting from designed laws, with the details, whether good or bad, left to the working out of what we may call chance."[11]

Darwin did not know the origin of new variations, but modern Darwinists believe that DNA mutations supply them. In 1970, French molecular biologist Jacques Monod said that with the discovery of DNA's structure and function, "and the understanding of the random physical basis of mutation that molecular biology has also provided, the mechanism of Darwinism is at last securely founded." Monod concluded, "Man has to understand that he is a mere accident."[12]

So living things may *look* as though they were designed, but if Darwinism is true then this is only an illusion. Oxford Darwinist Richard Dawkins even defines biology as "the study of complicated things that give the appearance of having been designed." Design is only an appearance, he believes, because "the evidence of evolution reveals a universe without design."[13]

Thus Darwinism consists of the following claims: (1) all living things are modified descendants of a common ancestor; (2) the principal mechanism of modification has been natural selection acting on undirected variations (originating in DNA mutations); and (3) unguided processes are sufficient to explain all features of living things—so design is an illusion.

Creation

Like evolution, "creation" has many meanings. In its broadest sense it simply means making something new; human beings create lots of things. Even when "creation" involves a being who transcends the natural world, it can have many meanings, from creating out of nothing to fashioning things from pre-existing materials.

With regard to living things, a creator might have made all species in their present forms in a single instant. Or a creator might have established universal laws and stepped back to let nature take its course. Between these two extremes there are many possible views.

As we saw above, Charles Darwin was "inclined to look at everything as resulting from designed laws." He also wrote in later editions of *The Origin of Species* that life may have "been originally breathed by the Creator into a few forms or into one."[14] If creation is defined to include the view that a creator designed the laws of the universe and intervened to make the first living cells, then even Darwin was a "creationist."

In the present controversy, however, the term is usually reserved for biblical creation. According to a literal reading of

The missing link is really a missing debate

"Define evolution as an issue of the history of the planet: as the way we try to understand change through time. The present is different from the past. Evolution happened; there is no debate within science as to whether it happened, and so on. I have used this approach at the college level."

—Eugenie C. Scott

the first verses of Genesis, God created the universe and living things in six days a little over six thousand years ago. But even Christians disagree over the interpretation of the Genesis "days." When Christian clergymen pioneered the modern study of geology in the early nineteenth century, many people interpreted Genesis to accommodate an old Earth. As a result, when Darwin published his theory in 1859 there was almost no opposition to it based on biblical chronology.[15]

What is now known as "young Earth creationism" did not rise to prominence until the middle of the twentieth century. Skirmishes between young Earth and old Earth creationists, and between both of these groups and the Darwinists, have been going on for decades, but they are not the source of the war declared by *Time* magazine in 2005.

The new war is not about evolution and creation, but about Darwinism and something called "intelligent design." What is it that Pat Shipman calls "horribly frightening" and Marshall Berman says "threatens all of science and society"?

Intelligent Design

According to the theory of intelligent design (ID), it is possible to infer from empirical evidence that some features of the natural world are best explained by an intelligent cause rather by than unguided processes. Although design arguments have a venerable history, the "ID movement," as it is sometimes called, is quite recent. It originated with the publication of several books between 1984 and 1992 and a small meeting organized by Berkeley law professor Phillip E. Johnson near Monterey, California, in 1993.[16]

Seven things are worth noting before we proceed further. First, the word "intelligent" emphasizes that "design" in this case is not just a pattern, but a pattern produced by a mind that conceives and executes a plan. Any natural causes involved are guided by intelligence. Writing a meaningful paragraph on a computer depends on various physiological, mechanical, and electronic processes, but without a mind directing them they would not produce the paragraph.

Second, ID is not a substitute for ignorance. If we don't know the cause of something that does not mean it was designed. When we make design inferences—and all of us make them every day—we do so on the basis of evidence; the more evidence, the more reliable the design inference.

Third, since intelligent design relies on scientific evidence rather than on Scripture or religious doctrines, it is not biblical creationism. Intelligent design makes no claims about biblical chronology, and biblical creationists have clearly distinguished their views from ID. A person does not even need to believe in God to infer intelligent design in nature; otherwise, prominent atheist Antony Flew could not have been persuaded that the evidence in nature points to design.[17]

Fourth, ID does not tell us the identity of the designer. Although most proponents of ID believe that the designer is the God of the Bible, they acknowledge that this belief goes beyond the scientific evidence. Thus ID is not the same as nineteenth-century natural theology, which reasoned from nature to the attributes of God. Instead, ID restricts itself to a simple question: does the evidence point to design in nature? The answer to this question—whether yes or no—carries implications for religious belief, but the question can be asked and answered without presupposing those implications.

Fifth, ID does not claim that design must be optimal; something may be designed even if it is flawed. When automobile manufacturers recall defective vehicles, they are showing that those vehicles were badly designed, not that they were undesigned.

Sixth, intelligent design is compatible with some aspects of Darwinian evolution. ID does not deny the reality of variation and natural selection; it just denies that those phenomena can accomplish all that Darwinists claim they can accomplish. ID does not maintain that all species were created in their present form; indeed, some ID advocates have no quarrel with the idea that all living things are descended from a common ancestor. ID challenges only the sufficiency of unguided natural processes and the Darwinian claim that design in living things is an illusion rather than a reality.

Finally, intelligent design can apply on two different levels. Design may be detectable in specific features of living things, but it may also be detectable in natural laws and the structure of the cosmos. Most people who consider themselves ID advocates maintain not only that design is empirically detectable in the cosmos as a whole, but also that some features of the natural world (such as the shapes of rocks at the base of a cliff) are not designed in the same sense that other features (such as the information in DNA) are designed.[18]

> ## Books That Started the Intelligent Design Movement
>
> *The Mystery of Life's Origin*, by Charles B. Thaxton, Walter L. Bradley, and Roger L. Olsen; Dallas, TX: Lewis and Stanley, 1984.
>
> *Evolution: A Theory in Crisis*, by Michael Denton; Bethesda, MD: Adler & Adler, 1985.
>
> *Darwin on Trial*, second edition, by Phillip E. Johnson; Downer's Grove, IL: InterVarsity Press, 1993.

War of the Words

The many meanings of "evolution" are frequently exploited by Darwinists to distract their critics. Eugenie Scott recommends: "Define evolution as an issue of the history of the planet: as the way we try to understand change through time. The present is different from the past. Evolution happened, there is no debate within science as to whether it happened, and so on...I have used this approach at the college level."[19]

Of course, no college student—indeed, no grade-school dropout—doubts that "the present is different from the past." Once Scott gets them nodding in agreement, she gradually introduces them to "The Big Idea" that all species—including monkeys and humans—are related through descent from a common ancestor. "Darwin called this 'descent with modification,' and it is still the best definition of evolution we can use."[20]

This tactic is called "equivocation"—changing the meaning of a term in the middle of an argument. Another tactic is to revise the history of science to discredit troublesome terminology. Harvard sociobiologist Edward O. Wilson recently claimed that the word "Darwinism" was coined by enemies of Darwin to make him look bad. "It's a rhetorical device to make evolution seem like a kind of faith, like 'Maoism'," said Wilson in *Newsweek* in November 2005. "Scientists," Wilson added, "don't call it Darwinism."[21]

Yet according to the *Oxford English Dictionary,* Thomas Henry Huxley (Darwin's most famous defender in Britain) used "Darwinism" in 1864 to describe Charles Darwin's theory. In 1876, Harvard botanist Asa Gray (who despite their disagreement over whether evolution was guided was Darwin's most ardent defender in America) published *Darwiniana: Essays and Reviews Pertaining to Darwinism,* and in 1889 natural selection's co-discoverer Alfred Russel Wallace published *Darwinism: An Exposition of the Theory of Natural Selection.* Two of Wilson's former Harvard colleagues, evolutionary biologists Ernst Mayr and Stephen Jay Gould, used the word extensively in their scientific writings, and recent science journals carry articles with titles such as "Darwinism and Immunology" and "The Integration of Darwinism and Evolutionary Morphology."[22]

Some people sugarcoat Darwinism to slip it down the throats of unsuspecting college students, while others falsely claim that the term is a creationist fabrication.

Books You're Not Supposed to Read

By Design or by Chance? The Growing Controversy on the Origins of Life in the Universe, by Denyse O'Leary; Minneapolis, MN: Augsburg Books, 2004.

Doubts About Darwin: A History of Intelligent Design, by Thomas Woodward; Grand Rapids, MI: Baker Books, 2004.

A Meaningful World: How the Arts and Sciences Reveal the Genius of Nature, by Benjamin Wiker and Jonathan Witt; Downers Grove, IL: InterVarsity Press, 2006.

Another source of confusion in the controversy is that intelligent design is often mis-defined. The most common definition of ID in the news media is that some aspects of nature are so complex they must have been designed. (Chapter Eight will explain in detail why this definition is incorrect.)

Wrong definitions such as this may be simply due to misunderstanding, but some Darwinists deliberately mis-define ID in order to discredit it. For example, philosopher Robert T. Pennock insists on calling ID "intelligent design creationism." Although

Websites You're Not Supposed to Visit

http://www.discovery.org/csc/

http://www.arn.org/

http://www.designinference.com/

http://www.uncommondescent.com/

http://www.iscid.org/

http://www.designorchance.com/

http://www.idthefuture.com/

(as we saw above) even Charles Darwin was a creationist by some definitions, calling ID "intelligent design creationism" in the context of the present controversy misleads people to confuse ID with biblical religion. For example, in 2005 science writer Matt Ridley called intelligent design "merely a dishonest attempt to repackage a literal interpretation of the Bible as science." University of Wisconsin (Madison) historian Ronald L. Numbers, an expert on creationism and a critic of intelligent design, says that it is inaccurate to call ID creationism—though it is the easiest way to discredit it.[23]

Despite all the honest confusion and dishonest misinformation, it isn't difficult to understand the issues in the war between Darwinism and intelligent design. We'll take them up one at a time in the chapters that follow.

According to the eminent Italian geneticist and Darwin critic Giuseppe Sermonti, "Darwinism is the politically correct of science." And according to Darwinists, intelligent design is a "horribly frightening" threat "to all of science and society." Since the book you now hold in your hands criticizes Darwinism and defends intelligent design, it is not only politically incorrect, but also politically dangerous.

Chapter Two

〰〰〰〰〰〰〰〰〰〰

WHAT THE FOSSIL RECORD *REALLY* SAYS

In 1998 and 1999, the U.S. National Academy of Sciences published two booklets defending Darwin's theory of evolution. According to the 1998 booklet, fossils provide the first of "several compelling lines of evidence that demonstrate beyond any reasonable doubt" that all living things are modified descendants of a common ancestor. The 1999 booklet claims that the theory has been "thoroughly tested and confirmed" by several categories of evidence—first of all the fossil record, which "provides consistent evidence of systematic change through time—of descent with modification."[1]

Many biology textbooks take the same line. In its section on "Evidence of Evolution," the widely used high school textbook *Prentice Hall Biology* gives the fossil record top billing and concludes: "By examining fossils from sequential layers of rock, one could view how a species had changed and produced different species over time."[2]

Fossils certainly prove that the Earth was once populated by creatures that are no longer with us. The fossil record also provides evidence that the history of life has passed through several stages, only the most recent of which includes us. For most pre-Darwinian geologists, the fossil record also counted against a literal six thousand year age for the Earth—though not against a recent creation of human beings.

Guess what?

- The most striking feature of the animal fossil record, the Cambrian explosion, turns Darwin's theory upside down.
- Darwinists claim they have found the missing links between land mammals and whales, but they admit that none of the links could be ancestors of the others.
- It is impossible, in principle, to show that any two fossils are genealogically related.

13

Do fossils also provide evidence for Darwin's theory that all living things are modified descendants of a common ancestor?

Darwin's Tree of Life

Imagine having a "chronoscope" that would enable you to peer back in time to the origin of the first animal—perhaps a primitive sponge. The sponge makes more sponges like itself, and (if Darwin's theory is true) after thousands of generations this sponge population splits into two different kinds of sponges, which we call separate species. After millions more generations and the origin of a few more species, some species become so different from each other that we group them into two genera (plural of "genus"). After countless more generations, the differences increase to the point where some genera are so different from each other that we divide them into two families. As differences continue to accumulate we eventually group various families into two or more orders, and various orders into two or more classes. Despite all the generations and all the differences, however, we might still have only sponges.

Then another major type of animal emerges—perhaps a jellyfish. This animal would be so radically different from the others that we wouldn't consider it just another class of sponge, but an entirely new category—a "phylum" (plural "phyla"). This pattern of gradual divergence from a common ancestor, with major differences arising only after a long accumulation of minor differences, is how Darwin envisioned evolution. "By the theory of natural selection," he wrote, "all living species have been connected with the parent-species of each genus, by differences not greater than we see between the natural and domestic varieties of the same species at the present day." According to Darwin's theory, if we could have observed the process of animal evolution "the number of intermediate and transitional links, between all living and extinct species, must have been inconceivably great."[3]

Those transitional links would have formed a branching pattern that Darwin called "the great Tree of Life." He illustrated this with a sketch in *The Origin of Species*. (Figure 1.)

If the "A" at the lower left in Darwin's illustration were the primitive sponge from which all other animals descended, most of the branches above it would still be sponges. The major differences—the phyla— would appear only at the top, after a long history of branching due to the accumulation of minor differences.

Modern biologists recognize several dozen animal phyla based on major differences in body plans. There are over a dozen phyla of worms alone, but there are even more striking differences between worms and

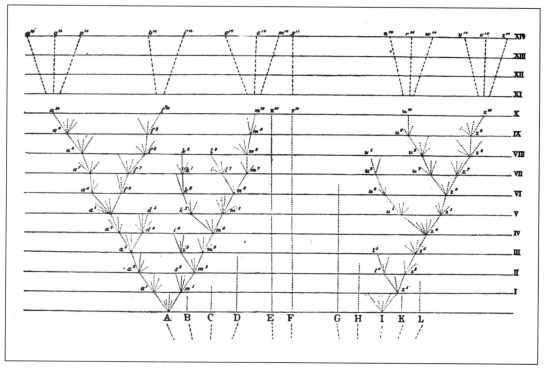

FIGURE 1. Darwin's Tree of Life

The oldest stage is at the bottom; the newest is at the top. The horizontal lines are separated by thousands or millions of generations.

mollusks (clams and octopuses), echinoderms (starfish and sea urchins), arthropods (lobsters and insects), and vertebrates (fish, amphibians, reptiles, birds, and mammals). If Darwin's theory were true, then these major differences should make their appearance at the top of his great Tree of Life.

But the fossil record shows exactly the opposite.

The Cambrian Explosion

When Darwin wrote *The Origin of Species*, the oldest known fossils were from a geological period known as the Cambrian, named after rocks in Cambria, Wales. But the Cambrian fossil record doesn't start with one or a few species that diverged gradually over millions of years into genera, then families, then orders, then classes, then phyla. Instead, most of the major animal phyla—and many of the major classes within them—appear together abruptly in the Cambrian, fully formed.

According to modern paleontologists James Valentine, Stanley Awramik, Philip Signor, and Peter Sadler, the appearance of the major animal phyla near the beginning of the Cambrian is "the single most spectacular phenomenon evident in the fossil record." The phenomenon is so dramatic that it has become known as the "Cambrian explosion," or "biology's Big Bang."[4]

Darwin was to some extent aware of this, and he called it a "serious" problem which "at present must remain inexplicable; and may be truly urged as a valid argument against the views here entertained." He discounted the problem by arguing that the innumerable transitional forms required by his theory must have existed, but they were either too small or too delicate to have been preserved in the fossil record. Many of Darwin's followers have relied on the same argument.[5]

In the past few decades, however, paleontologists have discovered microfossils of tiny bacteria in rocks estimated to be billions of years older than the Cambrian. Furthermore, detailed studies of fossils from the Cambrian explosion itself show that many of them were soft-bodied. According to Cambridge University paleontologist Simon Conway Morris: "These remarkable [Cambrian] fossils reveal not only their outlines but sometimes even internal organs such as the intestines or muscles." University of California–Los Angeles paleobiologist William Schopf wrote in 1994: "The long-held notion that Precambrian organisms must have been too small or too delicate to have been preserved in geological materials ... [is] now recognized as incorrect."[6]

Valentine and his colleagues agree that the Cambrian explosion "is real; it is too big to be masked by flaws in the fossil record." Indeed, as more fossils are discovered it becomes clear that the Cambrian explosion was "even more abrupt and extensive than previously envisioned."[7]

So the major phylum-level differences that Darwin predicted would appear last in the fossil record actually appeared first. Instead of proceeding from the bottom up, it seems that animal evolution—in the words of Valentine and his colleagues—"has by and large proceeded from the 'top down'." Paleontologist Harry Whittington, who pioneered the modern study of the Cambrian explosion in the Burgess Shale of Canada, wrote in 1985: "I look skeptically upon diagrams that show the branching diversity of animal life through time, and come down at the base to a single kind of animal ... Animals may have originated more than once, in different places and at different times."[8]

Nevertheless, most paleontologists—including Valentine and his colleagues—do not regard the Cambrian explosion as a refutation of Darwinian evolution. Indeed, Valentine recently reaffirmed his

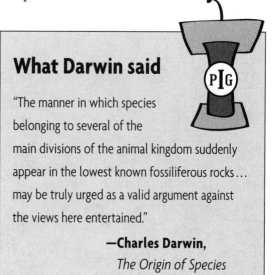

What Darwin said

"The manner in which species belonging to several of the main divisions of the animal kingdom suddenly appear in the lowest known fossiliferous rocks ... may be truly urged as a valid argument against the views here entertained."

—Charles Darwin,
The Origin of Species

conviction that "Darwin was correct in his conclusions that all living things have descended from a common ancestor."[9]

Whatever the source of the Darwinists' conviction may be, it cannot be the Cambrian fossil evidence. They can only affirm their belief in Darwinian evolution *in spite of* the Cambrian fossil record, not because of it.

A Whale of a Story

As far as fossil evidence is concerned, the Cambrian explosion may be Darwinism's worst-case scenario. In the interest of fairness and truth, however, we should also look at what many Darwinists consider to be their *best*-case scenario.

Land animals did not appear until long after the Cambrian period, and mammals did not appear until long after that. The fossil record shows only land mammals until after the extinction of the dinosaurs, but sometime after that whales appeared.

Fossils of dolphin-like dorudons and serpent-like basilosaurs were known even before Darwin published *The Origin of Species*. Both of these animals appear to have been fully aquatic. In 1983, a mammalian fossil skull was discovered in Pakistan that had some features suggestive of a whale, but the animal appears to have been a land-dweller.[10] Considering the large number of changes in anatomy and physiology that would be needed to turn a land mammal into a whale,

What Friday the 13th and the fossil record have in common

"The phrase 'the fossil record' sounds impressive and authoritative. As used by some persons it becomes, as intended, intimidating, taking on the aura of esoteric truth as expounded by an elite class of specialists. But what is it, really, this fossil record? Only data in search of interpretation. All claims to the contrary that I know, and I know of several, are so much superstition."

—Fossil expert
Gareth Nelson, 1978

many critics of Darwinism have argued that the absence of transitional forms between terrestrial and aquatic mammals is a serious problem for evolutionary theory.

According to the 1993 edition of the book *Of Pandas and People,* which criticizes Darwinism and defends intelligent design: "The extreme rarity of fossil transitional forms between the various types of plants, and the various types of animals, is a vexing problem for Darwinian thought." In particular, the book mentions whales: "Darwinists believe that whales evolved from a land mammal. The problem is that there are no clear transitional fossils linking land mammals to whales."[11]

The very next year, however, paleontologist Hans Thewissen and his colleagues reported the discovery in Pakistan of a fossil with characteristics intermediate between a land mammal and a whale. The animal had legs that would have enabled it to walk on land, like a modern sea lion, but it also had a long tail that would have enabled it to swim like a sea otter. Thewissen and his colleagues called their find *Ambulocetus natans,* or "swimming walking whale." A few months later, paleontologist Philip D. Gingerich and his colleagues discovered a slightly younger fossil in Pakistan that had some features intermediate between *Ambulocetus* and modern whales.[12]

Stephen Jay Gould called this "the sweetest series of transitional fossils an evolutionist could ever hope to find." Gould wrote: "This sequential discovery of picture-perfect intermediacy in the evolution of whales stands as a triumph in the history of paleontology. I cannot imagine a better tale for popular presentation of science or a more satisfying, and intellectually based, political victory over lingering creationist opposition."[13]

But no paleontologist worth his rocks—including Stephen Jay Gould— would claim that the series of whale fossils represents an actual lineage, because none of the animals could conceivably have given birth to any of the others. According to Berkeley paleontologist Kevin Padian, all of the fossil whales have "distinguishing characteristics, which they would have to lose in order to be considered direct ancestors of other known forms."[14] At best, each of the fossils represents a terminal side branch on the whales' tree of life. For example, in Figure 1, "the sweetest series of transitional fossils an evolutionist could ever hope to find" might occupy the branch-tips labeled s^2, i^4, k^8, and l^8. Darwinists acknowledge that not one of them would be in the "m" lineage leading to the modern whale, m^{10}.[15]

So the evidence from fossil whales is far better than the evidence from Cambrian fossils, but it still falls short of providing evidence for descent with modification. If one *assumes* that Darwin's theory is true, fossils showing features that appear to be intermediate between land mammals and whales can be placed in a series, but it is not a series of ancestors and descendants.

It turns out that the problem with fossils is not that transitional links are missing, but that fossil evidence *in principle* cannot provide evidence for descent with modification.

Bedtime Stories

In 1990, Ohio State University biologist Tim Berra published a book intended to refute critics of Darwinian evolution. To illustrate how the

How to put students to sleep: fossils

"No fossil is buried with its birth certificate. That, and the scarcity of fossils, means that it is effectively impossible to link fossils into chains of cause and effect in any valid way....To take a line of fossils and claim that they represent a lineage is not a scientific hypothesis that can be tested, but an assertion that carries the same validity as a bedtime story—amusing, perhaps even instructive, but not scientific."

—Evolutionary biologist **Henry Gee**, 1999

fossil record provides evidence for Darwin's theory of descent with modification, Berra used pictures of various models of Corvette automobiles. "If you compare a 1953 and a 1954 Corvette, side by side," he wrote, "then a 1954 and a 1955 model, and so on, the descent with modification is overwhelmingly obvious." But "descent" in Darwin's theory means biological continuity through the same reproductive processes we observe in living things today: fertilization, development and birth. Automobiles are made, not born. Corvettes actually prove the opposite of what Berra intended—namely, that a succession of similarities does *not,* in and of itself—provide evidence for biological descent with modification. Darwin critic Phillip E. Johnson calls this "Berra's Blunder."[16]

Even in the case of living things, which *do* show descent with modification within existing species, fossils cannot be used to establish ancestor-descendant relationships. Imagine finding two human skeletons in your back yard, one about thirty years older than the other. Was the older individual the parent of the younger? Without written genealogical records and identifying marks it is impossible to answer the question. And in this case we're dealing with two skeletons from the same species that are only a generation apart.

So even if we had a fossil representing every generation and every imaginable intermediate between land mammals and whales—if there were *no missing links whatsoever*, it would still be impossible *in principle* to establish ancestor-descendant relationships. At most, we could say that between land mammals and whales there are many intermediate steps; we could not conclude from the fossil record alone that any one step was descended from the one before it.

In 1978, fossil expert Gareth Nelson, of the American Museum of Natural History in New York, wrote: "The idea that one can go to the fossil record and expect to empirically recover an ancestor-descendant sequence, be it of species, genera, families, or whatever, has been, and continues to be, a pernicious illusion."[17]

Nature science writer Henry Gee doesn't doubt Darwinian evolution, but he candidly admits that we can't infer descent with modification from fossils. "No fossil is buried with its birth certificate," he wrote in 1999. "That, and the scarcity of fossils, means that it is effectively impossible to link fossils into chains of cause and effect in any valid way." According to Gee, we call new fossil discoveries missing links "as if the chain of ancestry and descent were a real object for our contemplation, and not what it really is: a completely human invention created after the fact, shaped to accord with human prejudices." He concluded: "To take a line of fossils and claim that they represent a lineage is not a scientific hypothesis that can be tested, but an assertion that carries the same validity as a bedtime story—amusing, perhaps even instructive, but not scientific."[18]

For many years, the California Academy of Sciences in San Francisco proudly featured a museum exhibit about Darwinism. Some of the fossils on display were so small that that magnifying glasses were positioned over them so curious schoolchildren could see them clearly. As visitors exited the museum they were treated to a "Hard Facts Wall," which showed an evolutionary tree of major animal groups. At each branch point in the tree—supposedly signifying the common ancestor of the branches above it—was a magnifying glass like those used elsewhere in the exhibit. Anyone who looked closely, however, could see that the magnifying glasses in this display had nothing under them. Visitors were expected to *imagine* common ancestors.[19]

Of course, one can *assume* that Darwin's theory is true, and then try to fit the fossil evidence into the picture suggested by that theory. There's nothing unreasonable about this—but let's state the reasoning up front: Theory rules, even without evidence. Fossils cannot provide evidence for descent with modification even when they're from the same

A Book You're Not Supposed to Read

Icons of Evolution: Why Much of What We Teach About Evolution Is Wrong, by Jonathan Wells; Washington, DC: Regnery Publishing, 2002.

species, much less when they're from entirely different species. Any claim to the contrary is just "a pernicious illusion" or "a bedtime story."

Want to Start a Barroom Fight?

Nevertheless, according to paleontologists Kevin Padian and Kenneth D. Angielczyk, it is "illogical to use the fossil record as a basis to assert ignorance of evolutionary patterns and processes." In a 1999 article arguing that the rarity of transitional fossils does not count as evidence against Darwinism, Padian and Angielczyk wrote: "Want to start a barroom fight? Ask another patron if he can produce proof of his unbroken patrilineal ancestry for the last four hundred years. Failing your challenge, the legitimacy of his birth is to be brought into question. At this insinuation, tables are overturned, convivial beverages spilled, and bottles fly. Not fair, claims the gentle reader. This goes beyond illogic to impoliteness, because you are not only placing on the other patron an unreasonable burden of proof, you are questioning his integrity if he fails. But isn't that what creationists do when they claim that our picture of evolution in the fossil record must be fraudulent because we have so many gaps between forms?"[20]

Yet Padian and Angielczyk have it exactly backwards.

Imagine this: A Berkeley professor walks into a bar and goes up to a guy who's peacefully sipping a beer. The professor looks down at the guy and declares with an air of authority: "You are the lineal descendant of a worm." The guy stands up, tempted to deck this bozo right then and there, but he's in a good mood and decides to play along. "Look," the guy says, "I've read about this Darwin stuff in the papers, but what makes you think you can tell me who I'm descended from? *I* don't even know anything about my great-great-grandparents, except that they were Irish. And here you are, claiming to know that one of their ancestors was a *worm*? Are you just trying to start a fight?"

"Look," says the professor, quoting comedian Lewis Black as his authority, "I'm right, and I don't have to argue this point any more. Fossils. Fossils. FOSSILS! I win."[21]

Chapter Three

≈≈≈≈≈≈≈≈≈≈≈≈≈≈≈

WHY YOU DIDN'T "EVOLVE" IN YOUR MOTHER'S WOMB

arwin knew that the fossil record was not good enough to establish his theory of descent with modification. He believed that his best evidence came not from fossils, but from embryos. "It seems to me," Darwin wrote in *The Origin of Species*, "the leading facts in embryology, which are second to none in importance, are explained on the principle of variations in the many descendants from some one ancient progenitor." And those leading facts, according to him, were that "the embryos of the most distinct species belonging to the same class are closely similar, but become, when fully developed, widely dissimilar." Darwin even believed that early embryos "show us, more or less completely, the condition of the progenitor of the whole group in its adult state." He considered this "by far the strongest single class of facts in favor of" his theory.[1]

Darwin's Strongest Evidence

Several decades before Darwin published *The Origin of Species*, German embryologist Karl Ernst von Baer had shown that the embryos of some vertebrates (animals with backbones) pass through a stage at which they look very much alike. The idea that vertebrates start out looking very similar as early embryos and then become progressively more different as

Guess what?

🦶 Darwin thought the strongest evidence for his theory was that vertebrate embryos are most similar in their earliest stages; the problem is, they're not.

🦶 Faked embryo drawings are still used in some modern biology textbooks as "evidence" for Darwin's theory.

🦶 Scientists have never been able to produce Darwinian evolution by mutating an embryo.

25

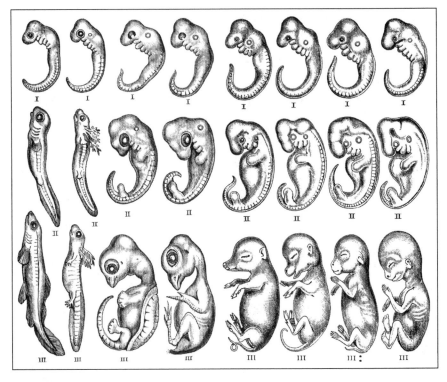

FIGURE 2.
Haeckel's Embryo Drawings

The embryos are (left to right) fish, salamander, tortoise, chick, hog, calf, rabbit, and human. The top row has been faked, and the four right-hand columns are all from the same order of mammals. This version of Haeckel's drawings is from Georges Romanes's 1892 book *Darwinism Illustrated*.

they develop into adults became known as "von Baer's law," though von Baer himself knew of many exceptions to it.[2]

Von Baer did not regard embryology as evidence for evolution. When Darwin proposed his theory, von Baer explicitly rejected the claim that the various classes of vertebrates (i.e., fishes, amphibians, reptiles, birds, and mammals) were descended from a common ancestor. According to historian of science Timothy Lenoir, von Baer criticized Darwinists for having "already accepted the Darwinian evolutionary hypothesis as true before they set to the task of observing embryos."[3]

Yet, in what historian of science Frederick Churchill calls "one of the ironies of nineteenth-century biology," von Baer's view "was confounded with and then transformed into" the evolutionary doctrine that the embryos of higher organisms pass through the adult forms of lower organisms in the course of their development. It was this evolutionary distortion of von Baer's work that Darwin considered the strongest evidence for his theory.[4]

In the 1860s, German Darwinist Ernst Haeckel (pronounced "heckle") made some drawings to illustrate this distorted view, and Darwin relied on the drawings in later editions of *The Origin of Species* and in *The Descent of Man* (1871). Haeckel's embryo drawings, in one form or another, have since been reprinted in millions of biology textbooks, which use them as evidence for Darwinism. (Figure 2)

But Haeckel faked his drawings. The embryos in the top row are not nearly as similar as he made them out to be; for more accurate drawings, see the next-to-bottom row in Figure 3. Furthermore, Haeckel was very selective in his choice of embryos. The four right-hand columns are all from the same order of mammals. Haeckel omitted embryos from the other two orders of mammals that include platypuses and kangaroos. He also omitted the two classes of vertebrates that include lampreys and sharks, and the order of amphibians that includes frogs— all of which look quite different from the groups portrayed here.[5]

Bending the Facts of Nature

Haeckel's fakery was exposed by his own contemporaries, who accused him of fraud, and it has been periodically re-exposed ever since. In 1997, British

Haeckel's embryo drawings are fakes

"It looks like it's turning out to be one of the most famous fakes in biology."

—Michael K. Richardson
Science, 1997

embryologist Michael K. Richardson and an international team of experts published photographs of actual embryos and compared them to Haeckel's drawings. When Richardson was interviewed by *Science* magazine, he said: "It looks like it's turning out to be one of the most famous fakes in biology."[6]

In 2000, evolutionary biologist Stephen Jay Gould called Haeckel's embryo drawings "fraudulent" and wrote: "We do, I think, have the right to be both astonished and ashamed by the century of mindless recycling that has led to the persistence of these drawings in a large number, if not a majority, of modern textbooks!" Yet the recycling continues. In 2004, Haeckel's embryo drawings were used as evidence for Darwinism in the tenth edition of Starr and Taggart's *Biology; The Unity and Diversity of Life*; in an early version of Raver's *Biology: Patterns and Processes of Life*; and in the third edition of Voet and Voet's *Biochemistry*.[7]

It seems that the "evidence" Haeckel supposedly provided for Darwin's theory is just too good to give up, and Darwinists continue to make excuses for his fakery. According to evolutionary biologist Jerry A. Coyne, Haeckel merely "doctored" his drawings. Anthropologist Eugenie C. Scott, of the militantly pro-Darwin National Center for Science Education

Turn the flock back: Darwin's Dodos

At the Tribeca Film Festival in April and May 2006, evolutionary biologist-turned-filmmaker Randy Olson premiered *Flock of Dodos*, a film that claims Haeckel's embryos haven't appeared in biology textbooks since 1914. Yet Olson knows that many recent textbooks *do* contain Haeckel's faked drawings. Although *Flock of Dodos* pretends to be a documentary, it is actually a pro-Darwin propaganda film.

(an organization we shall meet again and again throughout this book), says forgivingly that he "may have fudged his drawings somewhat." And textbook writer Douglas J. Futuyma notes euphemistically that Haeckel "did improve his drawings."[8]

In any other scientific field, people making excuses for fraud like this would probably be disgraced or drummed out of the profession.

Coyne, Scott, and Futuyma justify their defense of Haeckel by claiming that his faked drawings illustrate a deeper truth. Scott says "the basic point that's being illustrated by those drawings is still accurate." Coyne writes: "Embryos of different vertebrates tend to resemble one another in early stages, but diverge as development proceeds." And Futuyma claims that early vertebrate embryos "really are more similar, overall, than the animals are later in development."[9]

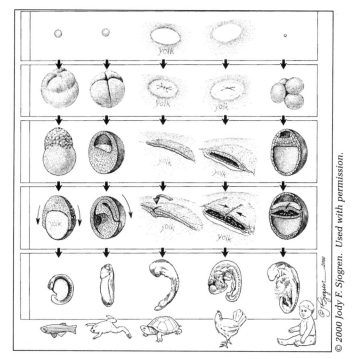

FIGURE 3.
Early Stages in Vertebrate Embryos

Drawings of early stages in the embryos of five classes of vertebrates. The stages are (top to bottom): fertilized egg; early cleavage; end of cleavage; gastrulation; and Haeckel's "first" stage. The fertilized eggs are drawn to scale relative to each other, while the scales of the succeeding stages are normalized to facilitate comparisons. The embryos are (left to right): bony fish (zebrafish), amphibian (frog), reptile (turtle), bird (chick) and mammal (human).

All of these statements are false.

When an animal egg is fertilized, it first goes through a process called "cleavage," during which it subdivides into hundreds or thousands of separate cells. At the end of cleavage, the cells begin to rearrange themselves in a process known as "'gastrulation," which is responsible for establishing the animal's shape and generating tissues and organs. According to British embryologist Lewis Wolpert, "it is not birth, marriage, or death, but gastrulation which is truly 'the important event in your life'."[10]

Yet only after cleavage and gastrulation does a vertebrate embryo reach the stage that Haeckel labeled the "first." If it were true (as Darwin and Haeckel claimed) that vertebrates are most similar in their earliest stages, then the various classes would be most similar during cleavage and gastrulation. Yet a survey of the five classes portrayed by Haeckel (bony fish, amphibian, reptile, bird, and mammal) reveals that this is not the case. The cleavage patterns in four of the five classes show some general similarities, but the pattern in mammals is radically different. In the gastrulation stage, a fish is very different from an amphibian, and both are very different from reptiles, birds, and mammals. This is certainly *not* a pattern in which the earliest stages are the most similar and later stages are more different. (Figure 3)

Like Haeckel's fakery, the dissimilarity of early vertebrate embryos was well known in the nineteenth century. Embryologist Adam Sedgwick pointed out in 1894 that the doctrine of early similarity and later difference is "not in accordance with the facts of development." Comparing a dogfish with a chicken, Sedgwick wrote: "There is no stage of development in which the unaided eye would fail to distinguish between them with ease." It is "not necessary to emphasize further these embryonic differences," Sedgwick continued, because "every embryologist knows that they exist and could bring forward innumerable instances of them. I need only say with regard to them that a species is distinct and distinguishable from its allies from the very earliest stages all through the development."[11]

Modern embryologists confirm this. Dartmouth College biologist William Ballard wrote in 1976 that it is "only by semantic tricks and subjective selection of evidence," by "bending the facts of nature," that one can argue that the cleavage and gastrulation stages of vertebrates "are more alike than their adults." And in 1987 embryologist Richard P. Elinson emphasized that the early developmental stages of frogs, chicks, and mice "are radically different."[12]

Darwinism Explains the Evidence—Away

So vertebrate embryos start out looking very different, then they become somewhat similar midway through development (though not as similar as Haeckel made them out to be) before diverging again. Embryologists call this pattern the "developmental hourglass." (Figure 4)[13]

Some recent biology textbooks have replaced Haeckel's faked drawings with photographs of

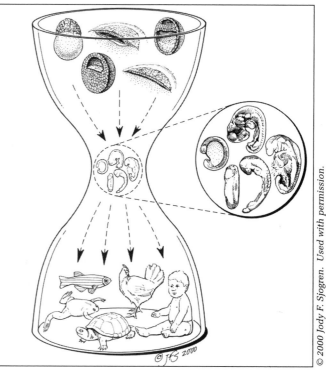

© 2000 Jody F. Sjogren. Used with permission.

FIGURE 4.
The Developmental Hourglass

Vertebrate embryos start out looking very different (top), then they converge somewhat in appearance midway through development (though not as much as Haeckel pretended), then they diverge again as they develop further (bottom).

actual embryos, but they continue to mislead students about the true pattern of vertebrate development. Miller and Levine's widely used high school textbook *Prentice Hall Biology* (2002) uses photographs to convince students that "in their early stages of development, chickens, turtles and rats look similar, providing evidence that they shared a common ancestor." Yet the book omits fish and frog embryos, which look very different, and its photographs of a turtle, chick, and rat are from the midpoint of development (the circle on the right-hand side of Figure 4) rather than an early stage.[14]

Darwin's strongest evidence, however, was the supposed similarity of embryos *in their earliest stages.* If embryos are the key to ancestry, as Darwin claimed, then the developmental hourglass pattern is more consistent with separate origins than common ancestry.

Instead of acknowledging the possibility that vertebrates might have different ancestors, Darwinists claim that the developmental hourglass pattern merely shows that early development can evolve easily. According to Duke University's Gregory Wray, "profound changes in developmental mechanisms can evolve quite rapidly," and Indiana University's Rudolf Raff argues that "the evolutionary freedom of early... stages is significant in providing novel developmental patterns."[15]

So when the strongest facts in favor of Darwin's theory turn out not to be facts at all, what do Darwinists do? Rather than question the theory, as scientists in other fields might, Darwinists simply declare that their theory is true anyway and use it to explain away those pesky embryos.

That raises my Haeckels

"We do, I think, have the right to be both astonished and ashamed by the century of mindless recycling that has led to the persistence of these drawings in a large number, if not a majority, of modern textbooks!"

—Stephen Jay Gould
Natural History, 2000

Evo-Devo to the Rescue?

Other than misrepresenting the pattern of vertebrate development, Darwinists made little use of embryology until the 1980s, when molecular biology made it possible to study the genes involved in embryo development. A new discipline emerged, called "evolutionary developmental biology," or "evo-devo" (pronounced ee-vo dee-vo).[16]

Ever since Darwinian evolution had been synthesized with Mendelian genetics in the 1930s, Darwinists had assumed that organisms are different because they have different genes. If genes control embryo development, then mutations in developmental genes would change the embryo, and the result would be evolution. In the 1980s, however, evo-devo biologists discovered to their surprise that radically different animals have very similar developmental genes. The principal developmental genes in mammals and insects are so similar that a gene from the former can replace its counterpart in the latter. For example, a gene needed for eye development in a mouse can induce eye development in a fruit fly embryo.[17]

The eyes induced in a fruit fly embryo by the mouse gene, however, are fruit fly eyes rather than mouse eyes. So whatever it is that makes a fly a fly or a mouse a mouse is *not* in the developmental genes they share. Those genes are non-specific switches, like the ignition switch in a vehicle. If we take an ignition switch from a Volvo convertible and put it in a Cessna jet, the plane doesn't turn into a car.

Darwinists argue that the remarkable similarity of developmental genes shared by different animals points to their common ancestry, though that doesn't explain how a relatively simple ancestral organism would have acquired all the developmental genes that are now found in its various and complex descendants. Even if the similarity of developmental genes were evidence for common ancestry, it would still constitute a paradox for neo-Darwinism. If genes control development, and radically different animals have similar developmental genes, then why are the animals so different? As Italian geneticist (and critic of Darwinism) Giuseppe Sermonti wrote in 2005, "Why is a fly not a horse?"[18]

A Fly Is a Fly Is a Fly

University of Wisconsin–Madison evo-devo biologist Sean B. Carroll argues that animals with similar developmental genes are different because they regulate their genes in different ways. Hitherto unexplored regions of DNA, he claims, regulate "when, where and how much of a gene's product is made,...and evolutionary changes within this regulatory DNA lead to the diversity of form."[19]

One of the most famous examples of the alleged power of changes in regulatory DNA is a fruit fly with an extra pair of wings. Normal fruit flies have one pair of wings and one pair of "balancers"—tiny appendages behind the wings that help to stabilize the insect in flight. In the 1970s, Cal Tech geneticist Edward B. Lewis discovered that by carefully breeding three mutant strains he was able to produce a fruit fly in which the balancers were transformed into a second pair of normal-looking wings. He also discovered that the mutations that produce a four-winged fruit fly are in the regulatory region of the *Ultrabithorax* gene, which produces a protein called Ubx. If the body segment behind the fly's normal wings produces Ubx, then balancers form; if mutations prevent it from producing Ubx, then normal-looking wings form.[20]

At first glance, this might seem to provide evidence for Carroll's claim that small developmental changes in regulatory DNA can produce large evolutionary changes in form. But the fruit fly is still a fruit fly. Furthermore, although the second pair of wings looks normal, it has no flight muscles. A four-winged fruit fly is like an airplane with a second pair of wings dangling uselessly from its tail. It has great difficulty flying or mating, so it can survive only in the laboratory. As evidence for evolution, a four-winged fruit fly is no better than a two-headed calf in a circus sideshow.[21]

A Book You're Not Supposed to Read

Why Is a Fly Not a Horse? by Giuseppe Sermonti; Seattle, WA: Discovery Institute Press, 2005.

In February 2002, a news release from the University of California–San Diego announced that biologist William McGinnis and his colleagues had discovered mutations that supposedly allowed shrimp-like animals, "with limbs on every segment of their bodies, to evolve 400 million years ago into a radically different body plan: the terrestrial six-legged insects." The news release boasted that this was "a landmark in evolutionary biology" because "it effectively answers a major criticism creationists had long leveled against evolution."[22]

The actual research, published in *Nature*, was far more modest. Shrimp embryos contain a version of the Ubx protein that does not inhibit leg formation, while a fruit fly contains no Ubx in its thorax (from which legs develop) but contains a version of Ubx in its abdomen that *does* inhibit leg formation. McGinnis and his colleagues showed that when Ubx from the abdomen of a fruit fly is inserted into the thorax of a fruit fly embryo, leg development is inhibited; but when Ubx from a shrimp abdomen is inserted into the thorax of a fruit fly embryo, normal fruit fly leg rudiments develop. They concluded that the Ubx in a shrimp-like ancestor must have mutated into the fruit fly version that now suppresses leg development.[23]

And maybe it did. But McGinnis and his colleagues did not reduce the number of legs in a shrimp, which is what supposedly happened in the course of evolution. Furthermore, even if they had shown how ancient shrimp lost a few legs, their experiment would not have even begun to explain how a water-dwelling shrimp-like animal could acquire the ability to breathe air and fly.

Evo-devo advocates have not given up, though. In a 2005 book titled *The Plausibility of Evolution*, Harvard's Marc W. Kirschner and Berkeley's John C. Gerhart argued that regulatory changes in DNA could make it "easy" for animals to evolve. They wrote that if such changes

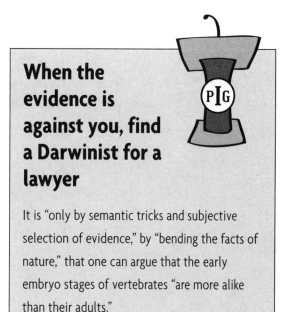

When the evidence is against you, find a Darwinist for a lawyer

It is "only by semantic tricks and subjective selection of evidence," by "bending the facts of nature," that one can argue that the early embryo stages of vertebrates "are more alike than their adults."

—William Ballard
BioScience, 1976

could be experimentally produced in a laboratory test animal then "doubters would have to admit" the plausibility of Darwinism. They concluded: "Such experiments are just now becoming feasible."[24]

Yet Kirschner and Gerhart cited only one such experiment. When a severe drought killed most of the finches on an island in the Galápagos in 1977, biologists observed that the survivors had, on average, slightly larger beaks. In 2004, a research team reported that Galápagos finches with larger beaks have more of a protein called Bmp4 in their embryos. When the researchers experimentally altered the amount of Bmp4 in chicken embryos, they found changes in the shapes of the embryos' beaks, though Bmp4 has other effects as well. The researchers concluded that changes in Bmp4 may have contributed to the beak changes in Galápagos finches, though they did not produce a breed of chickens with modified beaks, much less a new species of finch. (Neither did the 1977 drought: when the rains returned, the average beak size reverted to normal.)[25]

So evo-devo has provided us with evidence that regulatory changes in DNA can affect embryo development. But the results are always either trivial (Bmp4) or harmful (Ubx). Disabled fruit flies with extra wings or missing legs have taught us something about developmental genetics, but nothing about evolution. All of the evidence points to one conclusion: no matter what we do to a fruit fly embryo, there are only three possible outcomes—a normal fruit fly, a defective fruit fly, or a dead fruit fly. Not even a horsefly, much less a horse.

Chapter Four

❖❖❖❖❖❖❖❖❖❖❖❖❖

WHAT DO MOLECULES TELL US ABOUT OUR ANCESTORS?

Since fossils cannot *in principle* show us relationships of ancestry and descent, and since vertebrate embryos are not most similar in their earliest stages, the descent with modification of all living things from a common ancestor cannot be inferred from the fossil or embryo evidence. If Darwinism is science, it must be based on other evidence. Since the mid-twentieth century, Darwinists have increasingly turned to molecular biology to supply that evidence.

A 1999 booklet published by the U.S. National Academy of Sciences states: "The evidence for evolution from molecular biology is overwhelming and is growing quickly. In some cases, this molecular evidence makes it possible to go beyond the paleontological evidence. For example, it has long been postulated that whales descended from land mammals that had returned to the sea." Recent genetic comparisons "have confirmed this relationship."[1]

Ernst Haeckel coined the word "phylogeny" to refer to the evolutionary history of a group of organisms. If evolutionary relationships are inferred from comparisons of molecules—DNA, RNA, or proteins—the resulting pattern is called a "molecular phylogeny."

According to Douglas J. Futuyma's 2005 college textbook *Evolution*: "Molecular phylogenies support many of the relationships that have long been postulated from morphological [i.e., anatomical] data. These two

Guess what?

- Neither fossils nor embryos demostrate evolutionary relationships, so modern Darwinists construct hypotheses about the past from molecular data found only in living organisms.

- The only way to construct an evolutionary tree from molecules in living organisms is to *assume* Darwinism is true and then fit the data into a branching-tree pattern.

- Molecular studies have failed to produce a consistent evolutionary tree, and the more molecules scientists analyze, the more elusive the tree becomes.

data sets are entirely independent, . . . so their correspondence justifies confidence that the relationships are real."[2]

Yet there are serious inconsistencies between the molecular and morphological evidence. There are even serious inconsistencies between different analyses of the molecular data. Far from solving Darwinism's problems with the evidence, molecular phylogeny has aggravated them.

Molecular Phylogeny

Since we can't travel back in time, all hypotheses about past evolutionary relationships are inferred by comparing similarities and differences among organisms. With fossils, we're comparing things that lived in the past. With embryos, we're comparing things that are living now. In the 1960s, biologists Emile Zuckerkandl and Linus Pauling suggested that evolutionary histories could be reconstructed by comparing organisms at the molecular level. Except for a few rare DNA sequences that are several thousand years old, however, all of the molecules we have come from organisms that live in the present.[3]

All organisms, from bacteria to humans, contain DNA, RNA, and proteins. These molecules consist of sequences of subunits, but the exact sequences may differ from one organism to the next. Molecular phylogeny compares those sequences to infer evolutionary relationships. DNA, RNA, or proteins that differ by only a few subunits are presumed to be more closely related in evolutionary terms than those that differ by many subunits.

An attractive feature of molecular phylogeny is that it lends itself to quantitative measurement. It is more difficult to quantify the degree of similarity between the skeletons of a fish and a human (how does one assign a number to a shape?) than it is to quantify the number of subunits that are identical in two different DNA sequences.

Since all living cells use molecular factories called ribosomes to make proteins, molecular phylogeny has relied heavily on comparisons among the molecules in ribosomes. The most popular of these has been 18s rRNA. (The "18s" refers to its size, and "rRNA" is the abbreviation for "ribosomal RNA.")

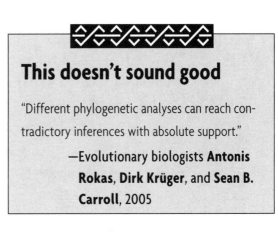

This doesn't sound good

"Different phylogenetic analyses can reach contradictory inferences with absolute support."

—Evolutionary biologists **Antonis Rokas, Dirk Krüger**, and **Sean B. Carroll**, 2005

In principle, it should be a simple matter to line up the 18s rRNA molecules from several different organisms, calculate the degree of similarity, and use those numbers to draw an evolutionary tree reflecting their ancestor-descendant relationships. In practice, however, this is not easy to do. First, it is not always clear how to line up two different molecules. Shifting a molecule one subunit relative to another molecule can radically alter the alignment and thus the result. Second, real biological molecules contain thousands of subunits, and if more than two molecules or more than two organisms are being compared, the numbers get very cumbersome. Many molecular biologists have spent their careers devising and refining sophisticated mathematical methods for constructing molecular phylogenies, and there is still considerable disagreement among such people over which methods are best.

With these things in mind, let's take a look at what molecular phylogeny has accomplished.

A Whale of a Story, Part 2

As we saw in Chapter Two, the whale series may be the Darwinists' best-case fossil scenario. Since the oldest specimens in the series consist of skulls and teeth that are similar to an extinct group of hyena-like mammals called mesonychians, University of Chicago evolutionary biologist Leigh Van Valen proposed in the 1960s that modern whales are descended from mesonychians. For several decades, that was the consensus view among paleontologists.[4]

In the 1990s, molecular studies suggested a very different picture. By comparing various molecules from whales with their counterparts in other living mammals, molecular evolutionists concluded that the closest living relatives of whales are hippopotamuses. Many paleontologists considered this anathema, since on morphological grounds hippos seem much more closely related to other even-toed hoofed mammals such as pigs and camels. On molecular grounds, however, the "whippo" hypothesis claimed that hippos are more closely related to whales than they are to other land mammals.[5]

Some evolutionary biologists remained skeptical. In 1999, John E. Heyning wrote in *Science*: "Previous experience suggests we should be cautious about wholeheartedly embracing such provocative hypotheses of relationships. More often than not, such controversial claims are found to be weakly supported or contradicted when scrutinized in more-detailed analyses." State University of New York biologists Maureen A. O'Leary and Jonathan H. Geisler examined 123 morphological characters from 10 living and 30 extinct species and concluded that whales are probably descended from mesonychians and that the "whippo" hypothesis is false.[6]

In 2001, scientists working with Philip D. Gingerich and J. G. M. Thewissen analyzed ankle bones in several newly discovered fossils and concluded that whales are more closely related to living even-

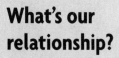

What's our relationship?

"Despite the comforting certainty of textbooks and 150 years of argument, the true relationships of the major groups (phyla) of animals remain contentious."

—Evolutionary biologists **Martin Jones** and **Mark Blaxter**, 2004

toed hoofed mammals than they are to mesonychians. The Gingerich team also concluded that it is "plausible" that hippos "may be the closest living relatives of whales," though the Thewissen group stopped short of claiming that hippos are closer to whales than they are to pigs or camels. In commentaries accompanying the reports, French biologist Christian de Muizon wrote that the new results "contradict the previous hypotheses of both paleontologists and molecular biologists," while American biologist Kenneth D. Rose pointed out that "substantial discrepancies remain" between the morphological and molecular evidence. In particular, removing mesonychians from the picture means the tooth and skull features that whales share with them had to be the result of "convergent evolution."[7]

"Convergent evolution" is a term Darwinists use for similarities that are not thought to result from common ancestry. But if similarities are the primary evidence for common ancestry, how can we know when they are *not* due to common ancestry? Similarities in fossils originally suggested that hippos are evolutionary sisters of pigs and camels but far removed from whales. Similarities in molecules now suggest that hippos are evolutionary sisters of whales but far removed from pigs and camels—and that the fossil similarities on which Darwinists originally relied were never evidence for common ancestry at all. If the original fossil similarities were not evidence for common ancestry, how do we know that the molecular similarities are? Why should we trust either hypothesis?

The answer matters, for two reasons. First, in order to understand how evolution works, it is important for biologists to know the route it took. To explain the origin of whales we need to know where they came from. Second, relationships are important. Imagine waking up one day and finding out that the people you thought were your parents and siblings are no more closely related to you than space aliens.

We all want to know our true family relationships. When it comes to molecular phylogeny, it turns out that we can't tell whether we are more closely related to insects than we are to worms.

Are We Closer to Insects or Worms?

The simplest and presumably most primitive animals are sponges and jellyfish. Other major animal groups, or phyla, have traditionally been divided into three broad categories based on morphological similarities. In this scheme, vertebrates (animals with backbones, such as humans) were considered more closely related to arthropods (crustaceans and insects) than to nematodes (tiny roundworms found in soil and marine sediments). Morphologically, we are evolutionary first cousins of insects, but only distant cousins of roundworms.

In 1997, Anne Marie Aguinaldo and her colleagues proposed a radical revision of animal relationships based on comparisons of their 18s rRNA. According to the revised phylogeny, we are no more closely related to insects than we are to roundworms.[8]

Obviously, Futuyma's claim that molecular phylogenies tend to support traditional morphological phylogenies is not true in this case. In fact, results from 28s RNA (a larger molecule also found in ribosomes) conflict with results from 18s rRNA. Even worse, 18s rRNA phylogenies differ from laboratory to laboratory. In 1999, evolutionary biologist Michael Lynch wrote: "Clarification of the phylogenetic relationships of the major animal phyla has been an elusive problem, with analyses based on different genes and even different analyses based on the same genes yielding a diversity of phylogenetic trees."[9]

Lynch was optimistic that with improved methods molecular phylogeny would clarify relationships among the animal phyla. Despite the continuing efforts of many researchers, however, conflicts among molecular phylogenies of the major animal groups have not just persisted, but grown.

In 2000, French molecular biologist André Adoutte and his colleagues affirmed their confidence in the new rRNA-based animal phylogeny. In 2002, however, Pennsylvania State University biologist Jaime E. Blair, along with his colleagues, compared over 100 protein alignments and concluded: "The grouping of nematodes with arthropods is an artifact that arose from the analysis of a single gene, 18s rRNA. The results presented here suggest caution in revising animal phylogeny from analyses of one or a few genes . . . . Our results indicate that insects (arthropods) are genetically and evolutionarily closer to humans (vertebrates) than to nematodes."[10]

In 2004, National Center for Biotechnology Information researcher Yuri I. Wolf and his colleagues analyzed over five hundred proteins using three different phylogenetic methods. They concluded that "the majority of the methods . . . grouped the fly with humans to the exclusion of nematodes." The following year, French biologist Hervé Philippe and colleagues analyzed 146 genes and 35 species representing 12 animal phyla. They concluded that their data grouped arthropods with nematodes to the exclusion of vertebrates.[11]

So round and round she goes, and where she stops, nobody knows. In a commentary accompanying Philippe's report, University of Edinburgh evolutionary biologists Martin Jones and Mark Blaxter wrote: "Despite the comforting certainty of textbooks and 150 years of argument, the true relationships of the major groups (phyla) of animals remain contentious." Although Jones and Blaxter favored Philippe's view, they predicted that the molecular tree of life "will sprout new shoots—and new controversies—very soon." Indeed, in December 2005, biologist Antonis Rokas and colleagues used two different methods to analyze fifty genes from seventeen animal groups. They noted that "different phylogenetic analyses can reach contradictory inferences with absolute support" and concluded that the evolutionary relationships among the phyla "remain unresolved."[12]

So the molecular phylogeny of the major animal phyla is a mess. When biologists use molecules to look for the trunk and root of Darwin's tree of life, the mess gets worse.

Uprooting the Tree of Life

According to Darwin, "all the organic beings which have ever lived on this earth may be descended from some one primordial form." In the 1980s, University of Illinois–Urbana microbiologist Carl R. Woese pioneered the use of 18s rRNA to construct an all-encompassing tree of life and identify the "universal common ancestor." By the late 1990s, serious problems had emerged. "When scientists started analyzing a variety of genes from different organisms," wrote University of California–Los Angeles molecular biologists James A. Lake, Ravi Jain, and Maria A. Rivera, they "found that their relationships to each other contradicted the evolutionary tree of life derived from rRNA analysis alone." According to French biologists Hervé Philippe and Patrick Forterre: "With more and more sequences available, it turned out that most protein phylogenies contradict each other as well as the rRNA tree."[13]

In 1998, Woese wrote: "No consistent organismal phylogeny has emerged from the many individual protein phylogenies so far produced." He concluded that primitive organisms acquired many of their genes and proteins, not by Darwinian descent with modification, but by "lateral gene transfer" from other organisms. "The universal ancestor," he wrote," is not an entity, a thing," but a community of complex molecules—a sort of primordial soup—from which different kinds of cells emerged independently.[14]

At about the same time, Dalhousie University evolutionary biologist W. Ford Doolittle concluded that lateral gene transfer among ancient organisms meant that molecular phylogeny might never be able to discover the "true tree" of life, not because it is using the wrong methods or the wrong genes, "but because the history of life cannot properly be represented as a tree." He concluded: "Perhaps it would be easier, and in the long run more productive, to abandon the attempt to force" the molecular data "into the mold provided by Darwin." Instead of a tree, Doolittle proposed "a web- or net-like pattern."[15]

Philippe and Forterre proposed methodological refinements that enabled them to re-root the universal tree, but the new root was not what everyone else was looking for. In the standard Darwinian scenario, life began with simple cells—cells without nuclei—but Philippe and Forterre rooted their phylogeny in a much more complex cell, one with a nucleus.[16]

The controversy over the universal tree of life continues. In 2002, Woese suggested that biology should go beyond Darwin's doctrine of common descent. In 2004, he wrote: "The root of the universal tree is an artifact resulting from forcing the evolutionary course into a tree representation when that representation is inappropriate." In 2004, Doolittle and his colleagues proposed replacing the tree of life with a net-like "synthesis of life," and in 2005 they recommended that "representations other than a tree should be investigated." Meanwhile, other scientists continue to defend the hypothesis that the universal ancestor existed but was complex rather than simple.[17]

In 2004, Rivera and Lake proposed yet another model. On the assumption that cells with nuclei originated when cells without nuclei fused together, Rivera and Lake inferred that "at the deepest levels . . . the tree of life is actually a ring of life." In an accompanying commentary, biologists

Barking up the wrong tree

"We cannot infer a unique tree of organisms from the pattern of relationships among genomes without making further assumptions about evolutionary processes that are just that: still-unproven assumptions. We have, for several decades, thought that our job was to uncover the structure of a Tree of Life, whose reality we need not question. But really, what we have been doing is testing Darwin's hypothesis that a tree is the appropriate representation of life's history, back to the beginning. Like any hypothesis, it could be false."

—Evolutionary biologist **W. Ford Doolittle**, 2005

William Martin and T. Martin Embley note that this "ring of life" is "at odds with the view of . . . simple Darwinian divergence."[18]

Whatever merits these hypotheses might have, one thing is clear: molecular phylogeny has failed, utterly and completely, to establish that universal common ancestry is true. The molecular evidence, like the fossil and embryo evidence, is plagued with inconsistencies, and Darwinism must be *assumed* in order to explain it; or, as is often the case, explain it *away*.

Reading the Entrails of Chickens

On the assumption that two organisms are descended from a common ancestor, their molecular differences are sometimes used to make quantitative estimates of how long ago they became separate lineages. Such estimates are calibrated against measured mutation rates in modern genes and dates derived from the fossil record. The problem is that different genes mutate at different rates, and dates derived from the fossil record are uncertain.

In a 2004 article in *Trends in Genetics*, evolutionary biologists Dan Graur and William Martin criticized some divergence dates that "were generated through improper methodology on the basis of a single calibration point" in the fossil record that was "both inaccurate and inexact—and in many instances inapplicable and irrelevant." Although all quantitative measurements in science include some error, much of the molecular dating literature has treated this calibration point as an errorless absolute, producing what Graur and Martin called an "illusion of precision" and "an exhaustive evolutionary

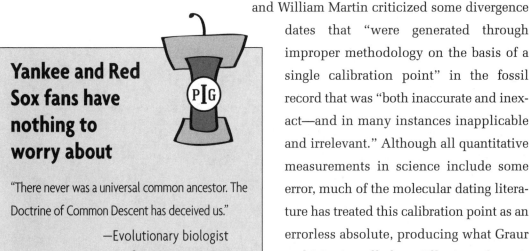

Yankee and Red Sox fans have nothing to worry about

"There never was a universal common ancestor. The Doctrine of Common Descent has deceived us."

—Evolutionary biologist **Carl R. Woese**, 2005

timeline that is enticing but totally imaginary." Graur and Martin concluded: "Unfortunately, no matter how great our thirst for glimpses of the past might be, mirages contain no water."

Graur and Martin's article was not a general critique of molecular phylogeny, or even of molecular dating. "On the contrary," they wrote, "molecular estimates of divergence times . . . are useful when based on solid statistical methodology and multiple fossil calibrations." But they were sufficiently frustrated with the imprecision they saw to title their article "Reading the Entrails of Chickens."[19]

In ancient Rome, some soothsayers used livers from sacrificed animals to foretell the future. Of course, the soothsayers were reading their own visions into the livers. Animal entrails, like tea leaves, are mute. But when it comes to providing evidence for Darwinian evolution, so are the molecules in living organisms.

Molecular inferences about whale ancestry conflict with morphological inferences; molecular phylogenies of the animal phyla conflict with each other; and inconsistencies in molecular phylogenies have persuaded many biologists to abandon the idea of a universal common ancestor altogether. Yet the main problem with molecular phylogenies is not that they conflict with the morphological evidence or with each other. Their main problem is that they *assume* the common ancestry they purport to prove. Descent with modification from a common ancestor has to be read into the molecules, which by themselves tell us nothing new or profound. As evidence for Darwinism, molecular phylogeny is no better than reading the entrails of chickens.

≈≈≈≈≈≈≈≈≈≈≈≈≈≈≈

Chapter Five

THE ULTIMATE MISSING LINK

I f all species (after the first) are descended with modification from some other species, then everything in Darwin's theory depends on the origin of new species from existing species—what evolutionary biologists call "speciation."

Changes within existing species are beside the point. "Darwin called his great work *On the Origin of Species*," wrote Harvard evolutionary biologist Ernst Mayr, "for he was fully conscious of the fact that the change from one species into another was the most fundamental problem of evolution."[1]

So speciation is Darwinism's most fundamental problem—the starting point for everything else in evolutionary theory. It is not an issue for intelligent design, which asserts only that some features of living things are best explained by an intelligent cause. Speciation is not an issue for biblical creationists, either, since Genesis states that God created "kinds," not individual species.

As a purely scientific matter, however, it is reasonable to ask: has speciation—the most fundamental process in Darwinian evolution—ever been observed?

Guess what?

- Although *all* species have supposedly descended from other species through selection and variation, no one has observed the origin of even *one* species by this process.
- Despite the title of his book, Darwin didn't solve the problem of the origin of species, and his followers are still looking for "evolution's smoking gun."
- Since fossils, embryos, and molecules don't demonstrate common ancestry and the "smoking gun" is still missing, the evidence for Darwinism is underwhelming, at best.

Evolution's Smoking Gun

Despite the title of his book, Darwin never solved the "mystery of mysteries," as he called the origin of species. In 1997, evolutionary biologist Keith Stewart Thomson wrote: "A matter of unfinished business for biologists is the identification of evolution's smoking gun," and "the smoking gun of evolution is speciation, not local adaptation and differentiation of populations." Before Darwin, the consensus was that species can vary only within certain limits; indeed, centuries of artificial selection had seemingly demonstrated such limits experimentally. "Darwin had to show that the limits could be broken," wrote Thomson, "so do we."[2]

Some Darwinists claim that they have found evolution's smoking gun by observing speciation in action. A 1987 article in *American Biology Teacher* cited "several instances of observed speciation that can... silence at least one common creationist argument against evolution." A pro-Darwin web site, Talk.Origins, subsequently posted a long essay titled "Observed Instances of Speciation," which was soon followed by "Some More Observed Speciation Events."[3]

Anyone who takes the time to plow through the references cited in these essays finds that most of the alleged instances of "observed" spe-

Biologists need the Nixon tape of June 23

"A matter of unfinished business for biologists is the identification of evolution's smoking gun.... The smoking gun of evolution is speciation, not local adaptation and differentiation of populations."

—Keith Stewart Thomson,
American Scientist, 1997

ciation are actually analyses of already existing species that are used to defend one or another hypothesis of how speciation occurs. For example, it seems likely that various species of finches on the Galápagos Islands are descended with modification from an original ancestor that came from the mainland of South America. It also seems likely that some similar fish species in African lakes had a recent origin. Biologists have proposed various hypotheses about how such species originated, and articles about them make up most of the alleged "observed instances of speciation" mentioned above. But analyzing existing species to test hypotheses about how they originated is a far cry from observing speciation in action.

There actually *are* some confirmed cases of observed speciation in plants—all of them due to an increase in the number of chromosomes, or "polyploidy." In the first decades of the twentieth century, Swedish scientist Arne Müntzing used two plant species to make a hybrid that underwent chromosome doubling to produce hempnettle, a member of the mint family that had already been found in nature. Polyploidy can also be physically or chemically induced without hybridization.[4]

Observed cases of speciation by polyploidy, however, are limited to flowering plants. According to evolutionary biologist Douglas J. Futuyma, polyploidy "does not confer major new morphological characteristics . . . [and] does not cause the evolution of new genera" or higher levels in the biological hierarchy. Darwinism depends on the splitting of one species into two, which then diverge and split and diverge and split, over and over again. Only this could produce the branching-tree pattern required by Darwinian evolution, in which all species are modified descendants of a common ancestor.[5]

So if biologists could start with one species and make it split into two by using variation and selection, they would be able to observe branching speciation, the cornerstone of Darwinism. They would have found evolution's smoking gun.

Speciation as a Research Program

Attempting to observe speciation is complicated by the fact that biologists have not been able to agree on a definition of "species," since no single definition fits every case. For example, a definition applicable to living, sexually reproducing organisms might make no sense when applied to fossils or bacteria. In their 2004 book *Speciation*, evolutionary biologists Jerry A. Coyne and H. Allen Orr point out that there are more than twenty-five definitions of "species." How can we choose among them? "Biologists want species concepts to be useful for some purpose," write Coyne and Orr, "but differ in what that purpose should be. We can think of at least five such goals." A species concept is useful, they explain, if it (1) helps biologists to classify organisms; (2) corresponds to the entities in nature; (3) helps us understand how those entities originate; (4) represents evolutionary history; and (5) applies to as many organisms as possible. Coyne and Orr acknowledge that "no species concept will accomplish even most of these purposes," but they "feel that, when deciding on a species concept, one should first identify the nature of one's 'species problem,' and then choose the concept best at solving that problem."

Like most other Darwinists, Coyne and Orr choose Ernst Mayr's "biological species concept" (BSC): "Species are groups of interbreeding natural populations that are reproductively isolated from other such groups." Why? "The most important advantage of the BSC is that it immediately suggests a research program to explain the existence of the entities it defines." Coyne and Orr "feel that it is less important to worry about species status than to recognize that the *process* of speciation involves acquiring reproductive barriers."[6]

In *The Origin of Species*, Darwin wrote: "According to my view, varieties are species in the process of formation, or are, as I have called them, incipient species." But how can we possibly know whether two varieties (or races) are in the process of becoming separate species? Saint Bernards and chihuahuas are two varieties that cannot interbreed naturally. Are they on their way to becoming separate species? How can we know? The Ainu people of northern Japan and the !Kung of southern Africa are separated not only geographically, linguistically, and culturally, but also (for all practical purposes) reproductively. Are they therefore "incipient species?"

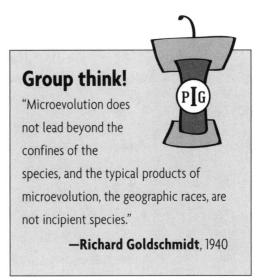

Group think!

"Microevolution does not lead beyond the confines of the species, and the typical products of microevolution, the geographic races, are not incipient species."

—**Richard Goldschmidt**, 1940

To people searching for evolution's smoking gun, reproductive isolation seems more promising than the many other criteria that distinguish separate species, because it can occur by degrees. But as the examples above demonstrate, partial reproductive isolation is no guarantee that two varieties are on their way to becoming separate species. "Incipient speciation" is really no more than a prediction, since by definition it refers to the production of two varieties that are thought to be en route to speciation.

With this in mind, let's look at some alleged instances of observed speciation.

Alleged Instances of Observed Speciation

The essays cited above in *American Biology Teacher* and on the Talk.Origins website list only five examples that might be construed as alleged instances of observed Darwinian speciation. First, from a single lab population of *Drosophila* (fruit flies) in 1962, J. M. Thoday and J. B Gibson bred only those with the highest and lowest number of bristles (the insect equivalent of hair). After twelve generations, the experiment produced two populations that not only differed in bristle number but also showed "strong though partial isolation." Not only did Thoday and Gibson *not* claim to have produced a new species, but other laboratories were unable to replicate their results.[7]

Second, in 1958 Theodosius Dobzhansky and Olga Pavlovsky started a laboratory population of fruit flies using a single female of the Llanos-A strain from Colombia. That same year, crosses between Llanos-A and several other strains produced fertile hybrids in the laboratory, but when tested again in 1963 similar crosses yielded sterile hybrids. In 1966, Dobzhansky and Pavlovsky concluded: "Llanos-A is a new race or incipient species having arisen in the laboratory at some time between 1958 and 1963." Coyne and Orr wrote in 2004, however, that Dobzhansky and Pavlovsky's result "may have been due to contamination of cultures by other subspecies." In any case, Dobzhansky and Pavlovsky reported only a "new race or incipient species," not a new species.[8]

Third, in 1964 biologists collected some marine worms from a population in Los Angeles Harbor and used them to start a lab colony. When biologists went back to the same location twenty-two years later, the original population had disappeared, so they collected worms from two other locations several miles away, and used them to start two new lab colonies. In 1989, researchers found that the two new colonies could interbreed with each other but not with the colony that had been started twenty-five years earlier. In 1992, James Weinberg and his colleagues called this an observed instance of "rapid speciation," based on the assumption that the original colony had "speciated in the laboratory, rather than before 1964." A few years later, however, tests performed by Weinberg and two others showed that the original population was "already a species different from" the two new colonies "at the time when it was originally sampled in 1964." No speciation had occurred.[9]

Fourth, in 1969 E. Paterniani reported an experiment on maize in which breeding was

No show

"Microevolution looks at adaptations that concern the survival of the fittest, not the arrival of the fittest....The origin of species—Darwin's problem—remains unsolved."

—**Scott Gilbert, John Optiz,**
and **Rudolf Raff,**
Developmental Biology,
1997

permitted only between individuals possessing two extremes of a particular trait. Paterniani noted "an almost complete reproductive isolation between two maize populations" but did not claim that a new species had been produced.[10]

Finally, in the 1980s William R. Rice and George W. Salt offered some fruit flies eight different habitats and bred only those that chose the two most extreme environments. Within thirty generations the flies had sorted themselves into two populations that did not interbreed, but Rice and Salt claimed only "incipient speciation that we believe to have occurred."[11]

So except for polyploidy in plants, which is not what Darwin's theory needs, there are no observed instances of the origin of species. As evolutionary biologists Lynn Margulis and Dorion Sagan wrote in 2002: "Speciation, whether in the remote Galápagos, in the laboratory cages of the drosophilosophers, or in the crowded sediments of the paleontologists, still has never been directly traced." Evolution's smoking gun is still missing.[12]

Microevolution and Macroevolution

According to Douglas J. Futuyma's 2005 college textbook *Evolution*, branching speciation "is the sine qua non of diversity" required to produce Darwin's branching tree of life. It "stands at the border between microevolution—the genetic changes within and among populations—and macroevolution"—the evolution of biological categories above the level of species. So one way of stating the importance of speciation is by distinguishing between "microevolution"—the uncontroversial changes within species that people observed long before Darwin—and "macroevolution"—the branching-tree pattern of evolution that is the essence of Darwinism.[13]

In 2005, Darwinist Gary Hurd claimed that the distinction between microevolution and macroevolution was just a creationist fabrication.

Asked to review proposed science standards requiring students to study evidence both for and against Darwinism, Hurd wrote to the Kansas State Board of Education: "I am confident that there are other qualified commentators who will have pointed out the absurdity of differentiating 'macro' and 'micro' evolution—terms which have no meaning outside of creationist polemics."[14]

Hurd represented himself as an expert on evolutionary biology, but the distinction between microevolution and macroevolution was first made by evolutionary biologists. In 1937, Theodosius Dobzhansky noted that there was no hard evidence to connect small-scale changes within existing species ("microevolution") to the origin of new species and the large-scale changes we see in the fossil record ("macroevolution"). Since "there is no way toward an understanding of the mechanisms of macroevolutionary changes, which require time on a geological scale, other than through a full comprehension of the microevolutionary processes observable within the span of a human lifetime," Dobzhansky concluded: "For this reason we are compelled at the present level of knowledge reluctantly to put a sign of equality between the mechanisms of macro- and microevolution, and proceeding on this assumption, to push our investigations as far ahead as this working hypothesis will permit."[15]

So Dobzhansky had to *assume* that microevolutionary processes are sufficient to account for macroevolution, and his assumption has been scientifically controversial ever since. In 1940, Berkeley geneticist Richard Goldschmidt published a book arguing that "the facts of microevolution do not suffice for an understanding of macroevolution." Goldschmidt concluded: "Microevolution does not lead beyond the confines of the species, and the typical products of microevolution, the geographic races, are not incipient species."[16]

In 1996, biologists Scott Gilbert, John Opitz, and Rudolf Raff wrote in the journal *Developmental Biology*: "Genetics might be adequate for explaining microevolution, but microevolutionary changes in gene frequency were not seen as able to turn a reptile into a mammal or to convert a fish into an amphibian. Microevolution looks at adaptations that concern the survival of the fittest, not the arrival of the fittest." They concluded: "The origin of species—Darwin's problem—remains unsolved." And in 2001, biologist Sean B. Carroll wrote in *Nature*: "A long-standing issue in evolutionary biology is whether the processes observable in extant populations and species (microevolution) are sufficient to account for the larger-scale changes evident over longer periods of life's history (macroevolution)."[17]

All of the biologists quoted above are Darwinists who accept the central tenets of evolutionary theory and believe that the controversy over microevolution and macroevolution will eventually be resolved within the framework of that theory. But their faith in a future resolution of the controversy does not alter the fact that the controversy exists—a century and a half after Darwin published *The Origin of Species*. And it exists because evolution's smoking gun is still missing.

Bacteria are conservatives

"Throughout 150 years of the science of bacteriology, there is no evidence that one species of bacteria has changed into another.... Since there is no evidence for species changes between the simplest forms of unicellular life, it is not surprising that there is no evidence for evolution...throughout the whole array of higher multicellular organisms."

—Bacteriologist **Alan H. Linton**, 2001

One Long Bluff

Darwin called *The Origin of Species* "one long argument." Its main point was that species change into other species through natural selection acting on unguided variations. In the century and a half since Darwin published his book, his followers claim to have amassed overwhelming evidence that the origin of species through Darwinian mechanisms is a scientific fact. In normal usage, however, a scientific fact is something we can observe, so their claim is false. Darwin may have given us one long argument, but his followers have given us one long bluff.

Although the bluffing sometimes occurs in the scientific literature itself, it usually comes from overzealous Darwinists eager to impress nonscientists. Some critics of Darwinism call this tactic "literature bluffing," and unfortunately it has become quite common.

For example, on June 9, 2004, the British Broadcasting Corporation (BBC) reported: "Scientists see new species born." But the scientific article on which the BBC based its report was about two existing species of fruit fly that hybridize to a limited extent. The scientists' actual conclusions? "Hybrid male sterility does not have a simple basis," and "earlier *Drosophila* speciation studies probably tell only a partial story."[18]

Like Gary Hurd, Darwinist Kenneth R. Miller wrote a review of the proposed 2005 Kansas science standards. Miller claimed that "the artificial distinction between micro- and macroevolution should be dropped," because "macroevolution has been observed repeatedly in nature." To prove his point, Miller cited a 2004 article in the *Proceedings of the National Academy of Sciences USA* about experiments performed on two existing fly species. But the authors of the article didn't even claim they had observed the formation of a new race, much less a new species. Miller was bluffing.[19]

When scientists are writing for other scientists, they tend to be more honest. In their 2004 book *Speciation*, Coyne and Orr noted that although selection in the laboratory can produce partial reproductive isolation or "incipient" speciation in ten to one hundred generations, "one does not expect full speciation to occur in so few generations."[20] If the origin of full species through Darwinian mechanisms takes too long to observe, then where is the "overwhelming evidence" for the sine qua non of Darwinism? Evidence that can never be found is not overwhelming; it's not evidence at all.

University of Bristol (England) bacteriologist Alan H. Linton went looking for direct evidence of speciation and concluded in 2001: "None exists in the literature claiming that one species has been shown to evolve into another. Bacteria, the simplest form of independent life, are ideal for this kind of study, with generation times of twenty to thirty minutes, and populations achieved after eighteen hours. But throughout 150 years of the science of bacteriology, there is no evidence that one species of bacteria has changed into another... Since there is no evidence for species changes between the simplest forms of unicellular life, it is not surprising that there is no evidence for evolution from prokaryotic [i.e., bacterial] to eukaryotic [i.e., plant and animal] cells, let alone throughout the whole array of higher multicellular organisms."[21]

So evolution's smoking gun is still missing. Darwinists claim that *all* species have descended from a common ancestor through variation and selection, but they can't point to a single observed instance in which even *one* species has originated in this way. Never in the field of science have so many based so much on so little.

The late Stephen Jay Gould—once called America's unofficial "evolutionist laureate"—loved baseball. If biology were a baseball game, the Darwinists would be celebrating right now in their locker room, surrounded by fawning journalists, as though they had just won the World Series. Outside, however, thousands of paying customers are still in the stands, scratching their heads. How could the Darwinists have won the game, the crowd wonders, when they never even made it to first base?

Chapter Six

❖❖❖❖❖❖❖❖❖❖❖❖❖❖

NOT EVEN A THEORY

In 1981, Harvard evolutionary biologist Stephen Jay Gould wrote: "Evolution is a theory. It is also a fact. And facts and theories are different things, not rungs in a hierarchy of increasing certainty. Facts are the world's data. Theories are structures of ideas that explain and interpret facts. Facts do not go away when scientists debate rival theories to explain them.... Humans evolved from ape-like ancestors, whether they did so by Darwin's proposed mechanism or by some other."[1]

In 1983, State University of New York evolutionary biologist Douglas J. Futuyma wrote that in biological evolution "we are dealing with two distinct questions. The first is the historical question of whether or not evolution has actually occurred. Have living forms actually descended by common ancestry from earlier forms? The second question is: if evolution has actually happened, what mechanisms have been responsible for it?" Futuyma concluded: "I consider the first question to have been resolved into fact, and the second question to fall into the category of theory."[2]

Textbook Controversies

In 1996, the Alabama State Board of Education voted to place labels inside biology textbooks stating (in part): "This textbook discusses evolution, a controversial theory some scientists present as a scientific

Guess what?

🐾 The claim that it is a "fact" that all living things evolved from a common ancestor is really just a way to protect Darwinism from critical analysis.

🐾 When a public school district decided to inform students that Darwinism is a "theory," not a fact, the Darwinists had the decision declared unconstitutional.

🐾 In the course of arguing their case, the Darwinists so exalted the meaning of "theory" that their doctrine no longer qualifies as one.

explanation for the origin of living things, such as plants, animals, and humans. No one was present when life first appeared on Earth. Therefore, any statement about life's origins should be considered as theory, not fact." The labels continued: "Evolution also refers to the unproven belief that random, undirected forces produced a world of living things."[3]

Darwinists were not pleased. In 2001, the U.S. Public Broadcasting System (PBS) televised a pro-Darwin series accompanied by a book, *Evolution: The Triumph of an Idea*. Gould wrote an introduction for the book that stated: "We should make a distinction, as Darwin explicitly did, between the simple *fact* of evolution—defined as the genealogical connection among all earthly organisms, based on their descent from a common ancestor... and *theories* (like Darwinian natural selection) that have been proposed to explain the causes of evolutionary change." But Gould then confused matters by adding: "Evolution substituted a naturalistic explanation of cold comfort for our former conviction that a benevolent deity fashioned us directly in his own image.... By taking the Darwinian 'cold bath,' and staring a factual reality in the face, we can finally abandon the cardinal false hope of the ages." So factuality, for Gould, applied not only to universal common ancestry but also to the naturalistic explanation that replaced God.[4]

In 2002, the Board of Education in Cobb County, Georgia, placed stickers inside their biology textbooks stating: "This textbook contains material on evolution. Evolution is a theory, not a fact, regarding the origin of living things. This material should be approached with an open mind, studied carefully, and critically considered." In August 2002, the pro-Darwin American Civil Liberties Union (ACLU) filed a lawsuit calling the stickers an unconstitutional establishment of religion.[5]

In January 2005, U.S. District Judge Clarence Cooper sided with the ACLU and ruled that "the distinction of evolution as a theory rather than a fact is the distinction that religiously motivated individuals have specifically asked school boards to make in the most recent anti-evolution movement.... Therefore, the sticker must be removed from all of the textbooks into which it has been placed."[6]

But it was Darwinists, not religiously motivated creationists, who made the distinction between evolution as fact and theory. As we saw in Chapter One, Darwinist Edward O. Wilson claimed that only creationists use the word Darwinism; and as we saw in Chapter Five, Darwinist Gary Hurd claimed that only creationists distinguish between microevolution and macroevolution. There's a pattern here, and we should have a word for it. Following the example of George Orwell's "Newspeak," how about "Darwinspeak"? When Darwinists object to having their own terminology used against them, Darwinspeak dismisses the terminology as a creationist invention.

Many Darwinists, however, continue to distinguish between evolution as fact and theory. In his 2005 college textbook *Evolution*, Futuyma wrote: "Evolution is a scientific fact. That is, the descent of all species, with modification, from common ancestors is a hypothesis that in the last 150 years or so has been supported by so much evidence, and has so successfully resisted all challenges, that it has become a fact." Evolution is also a "theory, the body of statements (about mutation, selection ... and so forth) that together account for the various changes that organisms have undergone." For Futuyma, theory "is a term of honor, reserved for principles ... that are well supported and provide a broad framework of explanation."[7]

But Darwinism is *not* well supported by the evidence—whether it is called a fact or a theory.

Underwhelming Evidence

When Alabama decided in 1996 to inform students that evolution is "theory, not fact," Oxford zoologist Richard Dawkins called the decision "a study in ignorance and dishonesty." He said: "A fact is a theory that is supported by all the evidence," and "the sheer weight of evidence, totally and utterly, sledgehammeringly, overwhelmingly strongly supports the conclusion that evolution is true." By evolution, Dawkins meant not only that "a tiny bacterium who lived in the sea was the ancestor of us all," but also that "macroevolution is nothing more than microevolution stretched out over a much greater time span." In other words, Dawkins was talking about Darwinism.[8]

So where is this "sledgehammeringly, overwhelmingly" strong evidence? As we saw in Chapter Two, the most striking phenomenon in the fossil record—the Cambrian explosion—does not fit Darwin's branching tree of life. In fact, fossils are unable *in principle* to provide evidence for ancestry and descent. In Chapter Three, we saw that Darwin's "strongest single class of facts"—early vertebrate embryos—shows the opposite of what he thought it showed. And no matter what we do to a fly embryo it remains a fly. Chapter Four illustrated that molecular phylogenies assume common ancestry rather than demonstrate it. And molecular phylogenies are notoriously inconsistent with each other. We might as well try to divine the past by reading the entrails of chickens.

As stated in Chapter Five, the most fundamental problem of evolution, the origin of species, remains unsolved. Despite centuries of artificial breeding and decades of laboratory experiments, no one has ever observed speciation through variation and selection. What Darwin claimed is true for *all* species has not been demonstrated for even *one* species. Trying to finesse the problem by calling minor changes within existing species "microevolution" is like trying to overcome obstacles to space travel by calling a toddler's first steps "microastronautics."

Sure, evolution is a fact in the sense that the present is different from the past; that the cosmos and living things have a history; that gene frequencies change from one generation

A Book You're Not Supposed to Read

Uncommon Dissent: Intellectuals Who Find Darwinism Unconvincing, edited by William A. Dembski; Wilmington, DE: ISI Books, 2004.

to the next; that individuals in the same species are related through descent with modification; and that variation and selection can produce changes within existing species. Evolution is also a collection of plausible but limited hypotheses—such as the possible Darwinian descent with modification of Galápagos finches.

But Darwinism claims a lot more than these few facts and limited hypotheses. Darwinism is the grand claim that *all* living things are modified descendants of a common ancestor, and that unguided natural processes alone—principally natural selection acting on random variations—have produced them.

The evidence that Darwinists cite for this grand claim is indeed overwhelming, if it's weighed by the pound, but it's largely irrelevant. Thousands of articles have been published in hundreds of science journals, but as evidence for Darwinism's grand claim they are just one long bluff. The Darwinists' "overwhelming evidence" has become a butt of jokes.[9]

Indeed, since their evidence is so *under*whelming, Darwinists now rely on "scientific consensus" to justify their position.

Science by Consensus

In 2003, University of Texas physicist and Nobel Prize winner Steven Weinberg testified before the Texas State Board of Education that although he was not a biologist he had "a good sense of how science works." "Science," he explained, "is what is generally accepted by scientists," and he assured the board "it is the theory of evolution through natural selection that has won general scientific acceptance."[10]

In 2005, the American Association of University Professors announced: "The theory of evolution is all but universally accepted in the community of scholars," and students should be taught "the overwhelming scientific consensus regarding evolution." At a 2006 pep rally for Darwinism in St. Louis, Eugenie Scott of the National Center for Science Education (NCSE) re-emphasized this overwhelming scientific consensus.[11]

Of course, nobody doubts that a large majority of professional biologists accept Darwinism. But appealing to majority opinion is a risky tactic in science, for three reasons. First, history shows that a "scientific consensus" is notoriously unreliable. The scientific consensus in 1600 was that the sun revolved around the Earth. The scientific consensus in 1750 was that things burn by giving off phlogiston. Indeed, the scientific consensus in 1900—four decades after *The Origin of Species*—was that Darwinism was false![12]

Second, majorities depend on who gets to vote. Polls have consistently shown that a significant majority of the American people reject Darwinism. Darwinists claim that the American people (who make up the most scientifically successful country in history) are scientifically illiterate, so their votes don't count. Even within the community of professional scientists, the situation is not as overwhelming as Darwinists would like us to believe. For many years, the NCSE claimed that no reputable scientists doubted Darwinism. When the Discovery Institute in Seattle started a "Dissent from Darwinism" list, which now includes more than five hundred Ph.D. scientists, the NCSE ridiculed it with a list of scientists named "Steve" (in honor of Stephen Jay Gould) who affirm Darwinism. By April 2006, "Project Steve" had attracted more than seven hundred signatures. But the NCSE's original claim had been that no qualified scientists doubt Darwinism—a claim that was decisively refuted long before the "Dissent from Darwinism" list reached the five hundred mark. "Project Steve" obscures the real point—that there is a significant and growing number of qualified scientists who dissent from Darwinism.[13]

The third problem with the "scientific consensus" approach is that once theories are accepted on the basis of majority opinion instead of evidence from nature, they become sociology rather than natural science. As an "overwhelming consensus" of professionals, Darwinism belongs in social studies classes instead of science classes.

So the scientific evidence for Darwinism is underwhelming, and history shows that a scientific consensus is unreliable. Why, then, do so many scientists put their faith in Darwinism?

Applied Materialistic Philosophy?

In *The Origin of Species*, Darwin repeatedly argued that his theory must be true because creation is false. For example: "Why, on the theory of Creation, should there be so much variety and so little real novelty?" Or: "Why should similar bones have been created to form the wing and the leg of a bat, used as they are for such totally different purposes?" Darwin's answer? These cases are "inexplicable...on the ordinary view of creation!"[14]

This is an odd way to defend a scientific theory. Would a geologist argue for continental drift by asking, "Why, on the theory of Creation, should the eastern contour of the Americas resemble the western contour of Europe and Africa?" Or would a physicist argue that unsupported dense objects fall to the ground because this is "inexplicable...on the ordinary view of creation!" Such arguments are silly. They have no legitimate place in natural science. Yet they are common in the literature defending Darwinism.

In a section on "Evidence for Evolution" in his 2005 college textbook *Evolution*, Futuyma wrote: "There are many examples, such as the eyes of vertebrates and cephalopod molluscs, in which functionally similar features actually differ profoundly in structure. Such differences are expected if structures are modified from features that differ in different ancestors, but are inconsistent with the notion that an omnipotent Creator, who should be able to adhere to an optimal design, provided them."[15] Where else in science would a statement about an omnipotent Creator constitute "evidence" for a theory?

Georgia State University historian Neal C. Gillespie pointed out in 1979 that "Darwin's theological defense of descent with modification" rested on his conception of the Creator, and that *The Origin of Species* "not only has numerous references to such a Creator, but theological arguments based on such a conception had some importance in its overall logic." Philosopher of biology Paul A. Nelson wrote in 1996: "A remarkable but little studied aspect of current evolutionary theory is the use by many biologists and philosophers of theological arguments for evolution."[16]

In 2001, biophysicist Cornelius G. Hunter concluded that the essence of Darwin's "one long argument" was that "evolution is true because divine creation is false." Darwin superimposed on nature his idea of "how God would go about creating the world" and found that the two did not match, "but the mismatch depends every bit as much on the theology as on the science." Thus Darwinism "was never scientific to begin with." According to Hunter, "our new metaphysical 'truths' are mistakenly viewed as the implications of evolution rather than the foundations of evolution."[17]

Gillespie had made a similar point in 1979: "It is sometimes said that Darwin converted the scientific world to evolution by showing them the process by which it had occurred . . . [but] it was more Darwin's insistence on totally natural explanations than on natural selection that won their adherence." The Darwinian revolution was primarily philosophical, and Darwin's philosophy was positivism. As Gillespie put it, positivism limits science to "the discovery of laws which reflected the operation of purely natural or 'secondary' causes." Furthermore, "there could be no out-of-bounds signs." Positivism assumes "that when sufficient natural or physical causes were not known they must nonetheless be assumed to exist to the exclusion of other causes." Gillespie concluded: "It was the prior success of positivism in science that assured the victory of evolution in biology."[18]

In other words, materialistic philosophy came first, and Darwinism followed. According to retired Berkeley law professor Phillip E. Johnson, Darwinism is essentially applied materialistic philosophy. Darwin's "theory of descent with modification made sense out of the pattern of natural relationships in a way that was acceptable to philosophical materialists." This is why Darwinists claim that universal common ancestry is a "fact" even though the evidence does not support it.[19]

"There is no real distinction," Johnson wrote, "between the 'fact' of evolution and Darwin's theory.... Ancestors give birth to descendants by the same reproductive process that we observe today, extended through millions of years. Like begets like, and so this process can only produce major transformations by accumulating the small differences that distinguish offspring from their parents.... All the basic elements of Darwinism are implied in the concept of ancestral descent." Thus "restating the theory as fact serves no purpose other than to protect it from falsification."[20]

So it's not evidence that makes Darwinism a "fact," but materialistic philosophy. In 1997, Harvard geneticist Richard C. Lewontin recounted how he and Carl Sagan had once defended Darwinism in a debate, then he explained: "We take the side of science *in spite* of the patent absurdity of some of its constructs... because we have a prior commitment, a commitment to materialism." Moreover, Lewontin continued, "that materialism is absolute, for we cannot allow a Divine Foot in the door."[21]

Too Good for Darwinism

When Judge Clarence Cooper ruled against Cobb County's textbook stickers in 2005, the Board of Education appealed the decision. Fifty-six professional scientific organizations joined together to oppose the appeal, stating: "The word 'theory' is reserved for our most well-substantiated and comprehensive explanations... that can incorporate facts, laws, inferences, and tested hypotheses."

A Book You're Not Supposed to Read

Darwin's God: Evolution and the Problem of Evil, by Cornelius G. Hunter; Grand Rapids, MI: Brazos Press, 2001.

Thus "scientific theories 'out-rank' facts by subsuming facts and laws within them."[22]

If theories "out-rank" facts, Darwinists should have been delighted when Cobb County called evolution "a theory, not a fact," but they feared that the public would mistake "theory" to mean a mere guess or speculation. The Federation of American Societies for Experimental Biology explained: "In science, a theory is a coherent explanation of natural phenomena based on direct observation or experimentation. Theories are logical, predictive, and testable. They are open to criticism and when shown to be false, they are modified or dismissed. Using this definition, evolution is categorized with other scientific theories such as gravity or atomic theory, which, like evolution, are universally accepted among scientists."[23]

As we have seen, however, the descent of all living things from a common ancestor through unguided natural processes is not "based on direct observation or experimentation"—nor can it be. Many of Darwinism's predictions about the fossil record, embryo patterns, and molecular comparisons have been "shown to be false"—yet it survives unmodified. And Darwinism is clearly not "universally accepted among scientists." Darwinism is not a fact. Indeed, as Darwinists themselves define the word, it is not even a theory.

Chapter Seven

∞∞∞∞∞∞∞∞∞∞∞∞∞

YOU'D THINK DARWIN CREATED THE INTERNET

Even though the evidence for Darwinism is so thin that it doesn't qualify as a "theory" by the Darwinists' own definition, evolutionary biologist Theodosius Dobzhansky wrote in 1973 that "nothing in biology makes sense" except in light of it.[1]

Modern Darwinists often make the same claim. In 1999, U.S. National Academy of Sciences president Bruce Alberts wrote: "The evolution of all the organisms that live on Earth today from ancestors that lived in the past is at the core of genetics, biochemistry, neurobiology, physiology, ecology, and other biological disciplines."[2]

Douglas J. Futuyma's 2005 college textbook *Evolution* declares: "Evolutionary biology is increasingly recognized for its usefulness: in fields as disparate as public health, agriculture, and computer science, the concepts, methods, and data of evolutionary biology make indispensable contributions to both basic and applied research . . . For anyone who envisions a career based on the life sciences—whether as physician or as biological researcher—an understanding of evolution is indispensable."[3]

"Nothing in biology makes sense" without Darwinism. It's "at the core of" all biological disciplines. It is "indispensable" to the life sciences. Or is it?

Agriculture and Genetics

The major advances in modern agriculture were not only pre-Darwinian but also predominantly political and mechanical. By 1800, the enclosure of fields had transformed European agriculture from subsistence-level feudalism to a more efficient system of individual ownership. Efficiency was further improved by the invention of seed drills, iron plows, and threshing machines—all before Darwin published *The Origin of Species* in 1859.[4]

Even after 1859, most agricultural advances had no connection to Darwin's theory. Better harrows, seeding equipment, and cultivators improved efficiency; the use of manure as fertilizer increased yields; silos prevented spoilage and improved the quality of feed; tractors were invented. Animal breeding and horticulture were well developed before 1800. Pre-Darwinian breeders knew the importance of selection, and many books about it were available in English by 1859.[5]

Modern agriculture *has* benefited from advances in biological science, but those advances have come largely from genetics, which originated with the work of Augustinian monk Gregor Mendel. Mendel found Darwin's theory unpersuasive.[6] The data he collected led him to conclude that heredity involves the transmission of stable factors that determine an organism's traits. Although the factors can be mixed and matched during reproduction, they remain discrete and unchanging from one generation to the next.

Darwin's view of heredity was quite different. He believed that every cell in an organism produces "gemmules" that transmit characteristics to the next generation in a blending process he called "pangenesis." The advantage of Darwin's view was that

Fairy obvious

"The Darwin-Wallace theory of evolution...is based on such flimsy assumptions, mainly of morphological-anatomical nature that it can hardly be called a theory....I would rather believe in fairies than in such wild speculation."

—Nobel laureate
Ernst Chain, 1972

gemmules could be changed by external conditions, or by use and disuse, and thus account for evolutionary change. The disadvantage of Darwin's view was that it was false.[7]

Mendel's theory of stable factors contradicted Darwin's theory of changeable gemmules. Thus, although Mendel's work was published in 1866, Darwinists totally ignored it for more than three decades. William Bateson, one of the scientists who "rediscovered" Mendelian genetics at the turn of the century, wrote that the cause for this lack of interest was "unquestionably to be found in that neglect of the experimental study of the problem of Species which supervened on the general acceptance of the Darwinian doctrines.... The question, it was imagined, had been answered and the debate ended."[8]

Even after 1900, Darwinists had little use for Mendel's theory. By the 1930s, however, the evidence had corroborated Mendelian genetics. Darwinists abandoned pangenesis and subsumed Mendelism in a "neo-Darwinian synthesis" that still dominates evolutionary biology.

Darwinist Bruce Alberts now claims that Darwinism is "at the core of genetics." Yet Mendel had no need for Darwin's hypothesis. How can Darwinism, which contributed nothing to the origin of genetics and resisted it for half a century, now be at its core? It is Darwinism that needs genetics, not genetics that needs Darwinism.

Medicine

Texas Tech University biology professor Michael Dini refuses to recommend for medical school any student who will not give a Darwinian account of the origin of the human species. According to Dini: "The central, unifying principle of biology is the theory of evolution, which includes both micro- and macroevolution, and which extends to ALL species. Someone who ignores the most important theory in biology cannot expect to properly practice in a field that is now so heavily based on biology."[9]

Yet modern medicine owes nothing to Darwinism. For one thing, mortality from infectious diseases in the West began declining before 1859, due in large part to public health measures such as the provision of sewage disposal systems and safe water supplies.[10] It also included personal hygiene, as the story of Hungarian obstetrician Ignác Semmelweis illustrates.

While working in an Austrian hospital in 1847, Semmelweis noticed that the death rate of mothers from puerperal fever was much higher in wards run by medical students than in wards run by midwives. He also noticed that the medical students would go directly from the morgue to the obstetric ward without washing their hands. By simply requiring the medical students to wash their hands in a chlorine solution, Semmelweis reduced mortality from 30 percent to less than 2 percent.[11]

The modern practice of immunization also originated without any help from Darwinism. Before 1800, smallpox was a serious and often fatal disease. In the 1790s, English physician Edward Jenner found that by vaccinating people with cowpox, a much milder disease, he could immunize them against smallpox. The worldwide elimination of smallpox in the twentieth century has been one of the most spectacular success stories in modern medicine. Yet Darwinism had nothing to do with it.[12]

Darwinists claim that their theory is needed to deal with viruses such as influenza that "evolve" from year to year. But the preparation of flu vaccines depends on techniques from the fields of virology, immunology, and biochemistry—not evolutionary biology.

The Discovery of Antibiotics

In 2001, the U.S. Public Broadcasting System (PBS) televised a pro-Darwin series accompanied by a book, *Evolution: The Triumph of an Idea*, which claimed: "The resistance that bacteria have to many antibiotics didn't just happen: it unfolded according to the principles of natural selection, as the bacteria with the best genes for fighting the drugs prospered. Without understanding evolution, a researcher has little hope of figuring out how to create new drugs and determine how they should be administered."[13]

Microbiologists generally use the word "antibiotic" to describe a substance produced by one microorganism that inhibits or kills other microorganisms. In this sense, English microbiologist Alexander Fleming discovered the first antibiotic: penicillin. Fleming noticed a culture dish of staph bacteria on which a mold spore had settled. Remarkably, there were no staph colonies around the mold, suggesting that the latter was producing a substance that killed or inhibited the staph bacteria. The mold was a species of *Penicillium* (another species of which is used to make blue cheese), and Fleming's training in microbiology enabled him to turn this serendipitous observation into a major medical breakthrough.[14]

Fleming published his discovery in 1929, but it wasn't until 1940 that chemists Howard Florey and Ernst Chain succeeded in purifying and concentrating the antibiotic enough to make it clinically useful. None of

Resistance to antibiotics

"The concept of the 'struggle for existence' has been applied to microbial interrelationships in nature in a manner comparable to the effects assigned by Darwin to higher forms of life. It has also been suggested that the ability of a microbe to produce an antibiotic substance enables it to survive in competition for space and for nutrients with other microbes. Such assumptions appear to be totally unjustified on the basis of existing knowledge.... All the discussion of a 'struggle for existence,' in which antibiotics are supposed to play a part, is merely a figment of the imagination, and an appeal to the melodramatic rather than the factual."

—Nobel laureate **Selman Waksman**, 1956

these three scientists saw any role for Darwinism in their work. In his Nobel Prize banquet speech in 1945, Fleming said he felt like a pawn "being moved about on the board of life by some superior power." Years later, Ernst Chain (who was Jewish) made it clear that he had no use for Darwinism at all. "The Darwin-Wallace theory of evolution," he said, "is based on such flimsy assumptions, mainly of morphological-anatomical nature that it can hardly be called a theory...I would rather believe in fairies than in such wild speculation."[15]

Penicillin is effective against many diseases, but not against tuberculosis, which causes millions of deaths worldwide every year. In 1944, microbiologist Selman Waksman and his research assistant Albert Schatz announced the discovery of streptomycin, and within a year it was being used to treat tuberculosis. Like the discoverers of penicillin, Waksman saw no role for Darwinism in the discovery of streptomycin. In 1956, he pointed out that the isolation, purification, and clinical application of antibiotics was highly artificial and had no counterpart in nature. Waksman concluded that the Darwinian assumption of a "struggle for existence" among microbes in nature is "totally unjustified."[16]

Subsequent research has supported this. Bacteria in the wild almost always live in communities with other microbes, including other species. Instead of reproducing as quickly as possible to out-compete their neighbors in a struggle for existence, they usually grow slowly and co-exist peacefully in the microscopic equivalent of a stable gated community.[17]

Resistance to Antibiotics

The clinical use of antibiotics creates a highly artificial situation. Antibiotic-producing microbes must be isolated from their natural surroundings and grown in pure culture with special nutrients. Then the antibiotic has to be purified and concentrated to a degree never seen in nature. When the antibiotic is finally administered to a patient, there is nothing "natural" about what follows. The greenhouses and livestock pens of domestic breeders are more natural than a hospital room or a doctor's office.

Occasionally, a few bacteria may survive antibiotic treatment. The survivors then multiply and continue the infection, against which the original antibiotic may be ineffective, and this can be a serious medical problem. Yet the process is not fundamentally different from domestic breeding, except that in domestic breeding it is the desirable ones that survive, while in antibiotic resistance it is the undesirable ones. Both cases involve human selection in an artificial situation, and neither case involves the origin of a new species. Tuberculosis bacteria that are resistant to antibiotics are still tuberculosis bacteria.

How do physicians deal with antibiotic resistance when it arises? They do not consult an evolutionary biologist. The two factors that contribute most to the emergence of antibiotic resistance are (1) improper use of antibiotics and (2) failure to isolate affected patients. In the 2003 edition

Some pioneers of modern biology who either preceded Darwin or rejected his theory

Andreas Vesalius (Anatomy), 1514–1564

William Harvey (Physiology), 1578–1657

Francesco Redi (Microbiology), 1626–1697

John Ray (Botany), 1627–1705

Anton van Leeuwenhoek (Microbiology), 1632–1723

Robert Hooke (Microbiology), 1635–1703

Carolus Linnaeus (Systematics), 1707–1778

Lazzaro Spallanzani (Reproductive Biology), 1729–1799

Caspar Friedrich Wolff (Embryology), 1734–1794

Georges Cuvier (Paleontology), 1769–1832

Karl Ernst von Baer (Embryology), 1792–1876

Richard Owen (Comparative Biology), 1804–1892

Louis Agassiz (Zoology), 1807–1873

Gregor Mendel (Genetics), 1822–1884

of *Principles and Practice of Pediatric Infectious Diseases*, Dr. Alan R. Hinman writes: "Major causes of antimicrobial resistance are the indiscriminate, inappropriate, inadequate, incomplete, and inconsistent use of antibiotics." These include using antibiotics as a preventive measure in animal feed; prescribing them for viral diseases such as the common cold, against which they are ineffective; and using inadequate dosages or discontinuing treatment too soon, thereby allowing some organisms to survive. According to the 2004 edition of Gorbach, Bartlett, and Blacklow's *Infectious Diseases,* the misuse of antibiotics "accounts for a significant proportion of the emerging resistance."[18]

Physicians deal with the second factor by relying on time-tested isolation procedures. They also deal with it by studying the mechanisms by which antibiotic resistance spreads from microbe to microbe—mechanisms involving gene transfer among organisms rather than Darwinian descent with modification. In the 2005 edition of Mandell, Douglas, and Bennett's *Principles and Practice of Infectious Diseases*, Drs. Steven M. Opal and Antone Medeiros write: "The best hope for the future is the development of a greater understanding of how microbial resistance spreads and the implementation of effective infection control strategies."[19]

New antibiotics are also being discovered without any help from evolutionary theory. Using microbiological screening procedures and organic chemistry, researchers at Johnson & Johnson have found a new antibiotic that in preliminary tests has proven effective against resistant strains of tuberculosis, and a group of Harvard chemists recently synthesized new forms of tetracycline (an antibiotic discovered in the 1950s) that promise to be effective against bacteria that are resistance to tetracycline itself. As Merck researchers Malcolm MacCoss and Thomas A. Baillie wrote in *Science* in 2004, a key element in new drug

A Book You're Not Supposed to Read

Getting the Facts Straight: A Viewer's Guide to PBS's Evolution. Seattle, WA: Discovery Institute Press, 2001.

discovery is "the continuing need for excellent synthetic chemists," not evolutionary biologists.[20]

Nothing in Biology?

Dobzhansky claimed that "nothing in biology makes sense except in the light of evolution." Yet most of the fundamental disciplines in modern biology were pioneered by scientists who lived before Darwin was born. Those pioneers include the sixteenth-century anatomist Andreas Vesalius, the sixteenth-century physiologist William Harvey, and the seventeenth-century botanist John Ray. They include the seventeenth-century founders of microbiology, Robert Hooke and Anton van Leeuwenhoek; the eighteenth-century founder of systematics, Carolus Linnaeus; and the eighteenth-century founder of modern embryology, Caspar Friedrich Wolff. Even paleontology, which Darwinists now treat as theirs, was founded before Darwin's birth by Georges Cuvier.

Several great pioneers in biology who lived to see the publication of *The Origin of Species* explicitly rejected Darwin's theory. These include embryologist Karl Ernst von Baer, comparative biologist Richard Owen, zoologist Louis Agassiz, and geneticist Gregor Mendel.[21]

One discipline deserves special mention: comparative biology. Darwinists sometimes claim that their theory helps us to understand what animals are most closely related to us and thus most likely to serve as models for human disease and drug testing. Such animals are identified on the basis of their genetic and biochemical similarities to us. This is just comparative biology at the level of genes and proteins. Linnaeus did comparative biology, yet he was a creationist who lived a century before Darwin; Owen and Agassiz did comparative biology, yet they rejected Darwin's theory. Mendel was no Darwinist, and Darwin was no biochemist. So comparative biology, like most other fields in biology, owes nothing to Darwinism.

In the final episode of PBS's television series, the narrator states that for decades after the 1925 Scopes trial "Darwin seemed to be locked out of America's public schools." When the Soviets launched Sputnik, the first man-made satellite, in 1957, according to the narrator, Darwin was restored to the curriculum and "long-neglected science programs were revived in America's classrooms." Yet during the supposedly benighted decades between 1925 and 1957, American schools produced more Nobel Prize winners than the rest of the world put together. And in physiology and medicine—the fields that should have been most stunted by a neglect of Darwinism—the U.S. produced fully twice as many Nobel laureates as all other countries combined. Obviously, biomedical science does just fine without Darwinism.[22]

Harvard biologist Marc W. Kirschner (Chapter Three) recently told a reporter for the *Boston Globe*: "Over the last one hundred years, almost all of biology has proceeded independent of evolution, except evolutionary biology itself." Although he lamented this situation, Kirschner acknowledged: "Molecular biology, biochemistry, physiology, have not taken evolution into account at all." Chemist Philip S. Skell, a member of the U.S. National Academy of Sciences, wrote in *The Scientist* that his "own research with antibiotics during World War II received no guidance from insights provided by Darwinian evolution." Skell had "recently asked more than seventy eminent researchers if they would have done their work differently if they had thought Darwin's theory was wrong. The responses were all the same: No." After reviewing the major biological discoveries of the twentieth century, Skell concluded: "I found that Darwin's theory had provided no discernible guidance, but was brought in, after the breakthroughs, as an interesting narrative gloss."[23]

Providing a narrative gloss may be interesting, but claiming credit for other people's achievements is theft. There should be a word for this particular form of intellectual larceny. Even though Darwin himself was not the culprit, when Darwinists steal credit for scientific breakthroughs to which they contributed nothing, the verb "to darwin" might be appropriate. Generations of breeders have been darwined. Mendel has been darwined. Jenner and Semmelweis have been darwined. Fleming, Florey, Chain, and Waksman have been darwined. So have the real pioneers of modern biology. They've all been darwined.

Taking Credit Where None Is Due

In 1999, then vice president Al Gore told CNN: "During my service in the United States Congress, I took the initiative in creating the Internet." Since what is now known as the Internet had been in the works since the early 1960s (when Gore was still a college student), Gore was obviously exaggerating. The candidate who claimed he "created the Internet" quickly became the butt of jokes from cartoonists and late-night talk-show hosts.[24]

What's Darwin got to do with it?

"I recently asked more than seventy eminent researchers if they would have done their work differently if they had thought Darwin's theory was wrong. The responses were all the same: no. I also examined the outstanding biodiscoveries of the past century: the discovery of the double helix; the characterization of the ribosome; the mapping of genomes; research on medications and drug reactions; improvements in food production and sanitation; the development of new surgeries; and others. I even queried biologists working in areas where one would expect the Darwinian paradigm to have most benefited research, such as the emergence of resistance to antibiotics and pesticides. Here, as elsewhere, I found that Darwin's theory had provided no discernible guidance, but was brought in, after the breakthroughs, as an interesting narrative gloss."

—National Academy of Sciences member **Philip S. Skell,** *The Scientist*, August 2005

Yet there was some truth to Gore's Internet claim. As a senator in 1988, he had introduced legislation to support a high-speed national computer network. Robert Kahn and Vinton Cerf, who designed the basic architecture and core protocols for the Internet, later wrote: "Al Gore was the first political leader to recognize the importance of the Internet and to promote and support its development." So although Gore would have been prudent to describe his role more modestly, his boast was not entirely without merit.[25]

Darwinists boast, too. But if nothing in biology makes sense except in the light of Darwinian evolution, how did it happen that most major biological disciplines were founded either before Darwin or by scientists who rejected his theory? Why do Darwinists claim that their hypothesis is indispensable for agriculture, when it was Darwin who needed farmers—not farmers who needed Darwin? How do Darwinists get away with claiming credit for Mendelian genetics, when Mendel doubted their theory and they ignored his work for decades? In what way is Darwinism indispensable to medicine, when the modern decline in infectious diseases resulted from public health measures and scientific disciplines that owe nothing to Darwin's theory?

Maybe it's time for late-night talk-show hosts to compare the way Darwinists take credit for biomedical science to the way Al Gore took credit for creating the Internet. Except that the comparison would be unfair. After all, Gore actually *did* have something to do with creating the Internet.

Chapter Eight

THE DESIGN REVOLUTION

erry passengers entering Victoria harbor in Canada are greeted by a bank covered with flowers that spell out "WELCOME TO VICTORIA" in large letters. Everyone who sees the greeting knows immediately that it was intelligently designed. In fact, all of us make design inferences every day. We wouldn't be able to function without them. But how do we do it? What sort of logic do we use?

Mathematician and philosopher William A. Dembski analyzed design inferences and concluded that they are not logically limited to the products of human activity. The same logic, he maintains, can be used to infer design in nature. As we saw in Chapter One, there's nothing radically new about this. Scientists, philosophers, and theologians have maintained for centuries that nature exhibits design. Yet Dembski's work has provoked a storm of controversy, primarily because it has brought new mathematical and philosophical rigor to the debate.

So what is intelligent design, and how do we detect it?

Design Inferences

The American news media typically define intelligent design as the idea that some things in nature are too complex to have originated by chance. This definition is wrong on two counts. First, nothing is "too

Guess what?

- We use ordinary logic to infer design in our daily lives, and intelligent design uses that same logic to infer design from evidence in nature.
- One way to infer design is to rule out explanations based on natural regularities or chance and to recognize patterns that have the hallmarks of intelligence.
- When Baylor University mathematician William Dembski organized a conference that included defenders as well as critics of intelligent design, Darwinists had him removed from his position.

complex" to have originated by chance. Complex things happen by chance all the time. When the Victoria harbor gardener is on vacation, the wind might cover the floral greeting with hundreds of dead leaves in a pattern that is more complex than the sign itself. Yet the pattern of dead leaves is not designed.

Second, we cannot infer design simply by showing that something did not originate by chance. The regular arrangement of sodium and chlorine atoms in a salt crystal is due not to chance, but to natural laws governing the packing of those particular atoms. A crystal is orderly, but it does not exhibit design—at least, not in the same sense that WELCOME TO VICTORIA exhibits design.

FIGURE 5.

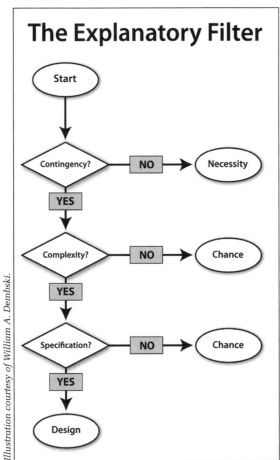

Illustration courtesy of William A. Dembski.

It is only when a pattern cannot plausibly be attributed either to chance or to natural regularity that we infer design. Dembski wrote in his 1998 book *The Design Inference*: "Whenever explaining an event, we must choose from three competing modes of explanation. These are *regularity*, *chance*, and *design*.... To attribute an event to design is to say that it cannot reasonably be referred to either regularity or chance." Thus "these two moves—ruling out regularity, and then ruling out chance—constitute the design inference."[1]

Dembski outlined the logic of design inferences in what he calls the explanatory filter. Dembski wrote: "Regularities are always the first line of defense. If we can explain by means of a regularity, chance and design are automatically precluded. Similarly, chance is always the second line of defense. If we can't explain by means of a regularity, but we can explain by means of chance, then design is automatically precluded. There is thus an order of priority to explanation. Within this order regularity has top priority, chance second, and design last."[2] (Figure 5)

The trickiest step in the process is eliminating chance. Something that is not complex could easily be due to chance. If several dozen letters of the alphabet were randomly lined up, we would not be surprised to find a two-letter word such as IT somewhere in the lineup. There are lots of two-letter English words, and finding one of them in a random string of letters would not be surprising. The second step in Dembski's explanatory filter attributes to chance anything that is not due to natural regularities but is not complex enough to warrant a design inference.

What if we were to pick several dozen letters and spaces randomly out of a hat and line them up, and they happened to form the sequence

WDLMNLT DTJBKW IRZREZLMQCO P

This sequence of twenty-eight words and spaces is extremely improbable. If we tried to obtain it again by drawing letters randomly from a hat, it would take us, on average, more than a trillion trillion trillion attempts. So the sequence is very complex, yet it is due to chance rather than design. The third step in the filter is intended to prevent us from inferring design in cases such as this by relying on a concept Dembski calls "specification."

A specification is a recognizable pattern that exists independently of the phenomenon we are analyzing. Dembski explains: "Plenty of highly

> ## Design mode
>
> "Whenever explaining an event, we must choose from three competing modes of explanation. These are *regularity, chance,* and *design.* . . . To attribute an event to design is to say that it cannot reasonably be referred to either regularity or chance."
>
> **—William A. Dembski,**
> *The Design Inference,* 1998

improbable events happen by chance all the time. The precise sequence of heads and tails on a long sequence of coin tosses ... can properly be explained by appeal to chance. It is only when the precise sequence of heads and tails has been recorded in advance ... that we begin to doubt whether these events occurred by chance. In other words, it is not just the sheer improbability of an event, but also the conformity of the event to a *pattern*, that leads us to look beyond chance to explain the event."[3]

We would not attribute the random sequence of twenty-eight letters and spaces above to design, but if we encountered twenty-eight letters and spaces that spelled

METHINKS IT IS LIKE A WEASEL

we would immediately infer that the sequence is designed. Not only is it extremely improbable (like the first sequence, it would occur on average less than one out of a trillion trillion trillion attempts), but it also corresponds to an independent pattern, namely, a line from Shakespeare's *Hamlet*. It is both *complex* and *specified*.

According to Dembski, such "specified complexity" is the hallmark of design: by rigorously eliminating regularity and chance, specified complexity leaves us with design as the only explanatory option.

No Free Lunch

Dembski concluded in *The Design Inference* that the explanatory filter "formalizes what we have been doing right along when we recognize intelligent agents." It is our universal human experience that whenever we encounter specified complexity it is a product of an intelligent agent. There is no free lunch. If specified complexity can be found in nature, then it, too, must be due to intelligent agency. As Dembski puts it: "The fundamental claim of intelligent design is straightforward and easily intelligible: namely, *there are natural systems that cannot be adequately explained in terms of undirected natural forces and that exhibit features which in any other circumstance we would attribute to intelligence."*[4]

Adding the word "intelligent" to "design" accomplishes two things. First, a design in the sense of mere pattern may be caused by natural forces (as waves cause ripple marks on a sandy beach) or even by an agent (as an absent-minded whittler causes knife marks on the end of a stick), yet not be a product of purpose. Adding "intelligent" to "design" makes it clear that the pattern in question results from the purposeful activity of a mind. Second, "intelligent" makes it clear that the design is not illusory, as Darwinists claim, but real.

Intelligence does not imply a violation of the laws of nature. When I chose to write this paragraph, I did not violate any law of nature. I used natural physiological processes to move my fingers and natural mechanical and electronic processes to record the words. Intelligent design does not suspend natural laws, it supplements them. This is why Dembski treats design as a third explanatory option, distinct from chance and necessity.

Nor does intelligence imply perfection. Some Darwinists criticize intelligent design on the grounds that some features of the natural world could have been made better. In other words, they argue that if something is not optimal, it is not designed. But things can be designed without being optimally designed. An automobile constructed in such a way that its fuel tank explodes whenever another vehicle bumps it from behind is badly designed, but it is designed nevertheless.[5]

The question remains whether specified complexity—and thus intelligent design—can be found in nature. Asking what features of living things may be designed is the subject of Chapters Nine and Ten, and asking the same question in relation to the cosmos as a whole is the subject of Chapter Eleven.

According to Darwinists, specified complexity in living things should not be attributed to intelligence because it can be explained by a combination of mutation and selection. Richard Dawkins instructed his computer to begin with the random sequence:

WDLMNLT DTJBKW IRZREZLMQCO P

and then to make random changes (mutations) in one letter or space at a time, selecting only those that matched a target line from *Hamlet*:

METHINKS IT IS LIKE A WEASEL

By randomly arranging letters and spaces, it would take on average more than a trillion trillion trillion attempts to produce this specific line, but Dawkins found that his "evolutionary algorithm" could produce it in only forty-three steps. He concluded that the "belief that Darwinian evolution is 'random' is not merely false. It is the exact opposite of the truth. Chance is a minor ingredient in the Darwinian recipe, but the most important ingredient is cumulative selection which is quintessentially *non*-random."[6]

Of course, Dawkins cheated by specifying the target in advance. He conceded as much, noting that at each step the phrase was compared to "a distant ideal target" even though "evolution has no long-term goal. There is no long-distance target." Nevertheless, Dawkins insisted that he could modify his computer model "to take account of this point" and still produce the desired outcome—though he didn't actually do so. "I don't know who," he wrote, "first pointed out that, given enough time, a monkey bashing away at random on a typewriter could produce all the works of Shakespeare."[7]

For now, it's sufficient to note that Dawkins's evolutionary algorithm didn't produce specified complexity—it had it all along. His computer exercise wasn't evidence for undirected Darwinian evolution, but for intelligent design. Dembski was right: there is no free lunch.

Dembski's explanatory filter is not the only possible way to formulate a design inference. Philosophers Timothy and Lydia McGrew, among others, agree that it is possible to infer design in nature, but they prefer an approach called an "Inference to the Best Explanation." According to philosopher Paul Thagard: "Inference to a scientific theory is not only a matter of the relation of the theory to the evidence, but must also take into account the relation of competing theories to the evidence. Inference is a matter of choosing among alternative theories, and we choose according to which one provides the best explanation."[8]

Dembski's Dangerous Idea

When New Testament scholar Robert Sloan became president of Baylor University in 1995, he saw intelligent design as one way to bridge the widening gap between the secular academy and Baylor's Christian roots. In 1998, he invited Dembski to set up a research center focusing on the controversial issue of intelligent design. As Baylor's provost said: "This was an opportunity to reaffirm that Baylor is a university where controversial issues can be discussed."[9]

Dembski signed a five-year contract, established the Michael Polanyi Center (named after an eminent chemist and philosopher of science), and began planning a major conference at Baylor for April 2000, titled "The Nature of Nature." The conference would focus on the question:

That's easy

"The fundamental claim of intelligent design is straightforward and easily intelligible: namely, *there are natural systems that cannot be adequately explained in terms of undirected natural forces and that exhibit features which in any other circumstance we would attribute to intelligence.*"

—William A. Dembski, *The Design Revolution,* 2004

"Is the universe self-contained or does it require something beyond itself to explain its existence and internal function?" The invitation list included Christians and atheists, prominent philosophers and scientists. Meanwhile some Baylor faculty were becoming upset about the center's very existence. So was an atheist philosopher in another state, Barbara Forrest.

Forrest is a professor at Southeastern Louisiana University. She is also a member of the New Orleans Secular Humanist Association, an organization that opposes efforts "to explain the world in supernatural terms." A month before the conference was scheduled to begin, Forrest sent a five-page message to many of the invitees warning them that they were being "lured" by "creationists with a religious/political agenda" to a meeting that would lend "undeserved academic legitimacy" to a movement bent on "the destruction of evolutionary theory."[10]

The conference was held anyway, and by all accounts it was a great success. Three hundred and fifty scholars from around the world, including two Nobel laureates, spent four days listening to lectures and participating in lively academic discussions.[11] Critics of ID were well represented—though not by the Baylor biology faculty, which boycotted the event. Nevertheless, a few days after the conference the Baylor faculty senate voted to recommend that President Sloan close the Michael Polanyi Center.

A few weeks later, some U.S. congressmen attended an informational meeting on intelligent design in Washington, D.C. Alarmed, eight Baylor Darwinists wrote to Congressman Mark Souder to complain: "We and other mainstream scientists refer to it as intelligent design creationism. Some have referred to it as 'creeping creationism'." The letter writers criticized intelligent design for disparaging "the Darwinian and materialistic worldview" to which (they claimed) the

A Book You're Not Supposed to Read

The Design Revolution: Answering the Toughest Questions About Intelligent Design, by William A. Dembski; Downers Grove, IL: InterVarsity Press, 2004.

American people owe the conquest of racial and gender inequality, infant mortality, smallpox, and polio.

The letter backfired when Souder blasted the Baylor Darwinists on the floor of the House of Representatives. "Ideological bias has no place in science," Souder said, "and many of us in Congress do not want the government to be party to it." He concluded, "As the Congress, it might be wise for us to question whether the legitimate authority of science over scientific matters is being misused by persons who wish to identify science with a philosophy they prefer. Does the scientific community really welcome new ideas and dissent, or does it merely pay lip service to them while imposing a materialist orthodoxy?"[12]

At first, President Sloan resisted faculty pressure to close the Michael Polanyi Center. "It's rather ironic," he said, "that people in the scientific community whose rights had to be protected in the face of ideological pressure" from fundamentalists "now appear to be suppressing others." Sloan concluded that the faculty's position "borders on McCarthyism." Nevertheless, the faculty forced Sloan to appoint a committee to review the status of the Michael Polanyi Center. Since the faculty made sure that the committee was stacked with biologists hostile to Dembski, it came as no surprise when it reaffirmed the faculty senate's recommendation that the Center be closed. Dembski was removed from his position (though he stayed on at Baylor as an independent researcher until his contract expired), and eventually even Sloan lost his job.[13]

The Baylor lesson is clear: Darwinists will not tolerate any open discussion of intelligent design. When they cannot crush it in a balanced academic forum, they resort to intimidation and mob rule. In the following chapters we shall see this pattern repeated over and over again. To anyone hoping for a career in the current politically correct academic establishment, intelligent design is professionally dangerous. And no one knows this better than Dembski himself.

Monkeys Typing Shakespeare

Dawkins didn't know who originated the idea that a monkey, given enough time, could type the works of Shakespeare. Actually, the idea originated with French mathematician Émile Borel, who pointed out in 1913 how unlikely it would be that a million monkeys typing ten hours a day would produce exactly all the books in the world's libraries. Borel used this to illustrate the extreme *un*likelihood of certain events, but Darwinists such as Dawkins now use the image of typing monkeys to illustrate the opposite: the likelihood that random processes can produce information. According to the "infinite monkey theorem," an infinite number of typing monkeys (or a finite number of monkeys typing for an infinite time) will eventually produce the works of William Shakespeare.[14]

In theory, an infinite number of monkeys with an infinite amount of time *would* write the works of Shakespeare—mixed in with an infinite amount of utter nonsense. But the real world isn't infinite. On July 1, 2003, a Monkey Shakespeare Simulator was posted on the Internet. The simulator does not use real monkeys, of course; it is a computerized random letter generator in which each "monkey" types one letter per second and the number of monkeys is continually increasing. The simulator compares its output with the works of Shakespeare and reports matches—though it waits for such matches to appear on their own instead of selecting for a target sequence, as Dawkins's computer did.

After a year and a half, the longest match produced by the Monkey Shakespeare Simulator was twenty-four letters

Can I infer from what you are saying that . . .

"Inference to a scientific theory is not only a matter of the relation of the theory to the evidence, but must also take into account the relation of competing theories to the evidence. Inference is a matter of choosing among alternative theories, and we choose according to which one provides the best explanation."

—Paul Thagard,
The Journal of Philosophy, 1978

from a line in *The Second Part of King Henry IV*, which took the equivalent of 2,738 trillion trillion trillion monkey-years to produce. (A year later, the record had been extended to over thirty letters, which took trillions and trillions more monkey-years to produce.)[15]

So the universe isn't big enough or old enough to hold all the "monkeys" it would take to type even one of Shakespeare's sonnets—much less his collected works. And real monkeys don't type a letter every second without stopping. What happens when we use real monkeys?

In 2003, some lecturers and students from Plymouth University in southwest England decided to find out. They took a computer to the Paignton Zoo, about sixty miles away, and put it in a monkey enclosure housing six crested macaques. Then they waited. At first, a male monkey started bashing on the computer with a rock. "Another thing they were interested in was in defecating and urinating all over the keyboard," added Mike Phillips, who runs the university's Institute of Digital Arts and Technologies. After a month, the monkeys had produced the equivalent of five typed pages, consisting almost entirely of the letter "S." They failed to produce anything remotely resembling a word.[16]

In the process, the monkeys received lots of free lunches. But they produced no specified complexity. No Shakespeare. No "WELCOME TO VICTORIA." Not a single word.

Chapter Nine

❊❊❊❊❊❊❊❊❊❊❊❊❊

THE SECRET OF LIFE

At the Cavendish Laboratory in Cambridge, England, in 1953, James Watson and Francis Crick had just spent months trying to figure out the molecular structure of DNA. When they finally succeeded one Saturday morning in February, they went to celebrate over drinks at the nearby Eagle pub. To those around them, Crick announced: "We have discovered the secret of life!"[1]

Watson and Crick's breakthrough led to further discoveries showing how DNA encodes the genetic information that prescribes the protein building blocks of living cells, how that information is copied and transmitted to future generations, and how molecular accidents—mutations— might introduce changes in the genetic information to provide the raw materials for evolution.

In 1970, French molecular biologist Jacques Monod again called DNA the secret of life. He said that with the new understanding of its structure and function, "and the understanding of the random physical basis of mutation that molecular biology has also provided, the mechanism of Darwinism is at last securely founded." Monod concluded: "Man has to understand that he is a mere accident."[2]

But Monod was wrong. The secret of life is not the physical DNA molecule, but the information it carries.

Guess what?

- The secret of DNA's success is that it carries information like that of a computer program, but far more advanced.
- Since experience shows that intelligence is the only presently acting cause of information, we can infer that intelligence is the best explanation for the information in DNA.
- When philospher Stephen Meyer published this inference in a biology journal, with supporting scientific evidence, Darwinists tried to ruin the career of the journal's editor.

DNA: the Molecule

Living cells contain several different classes of large organic molecules, including proteins, DNA, RNA, lipids, and complex carbohydrates. With the development of Mendelian genetics in the early twentieth century, the cellular structures that seemed to be the likeliest carriers of genes were the chromosomes, which were known to consist largely of DNA. But DNA was thought to consist of simple repeated sets of its four subunits: A, T, C, and G. Most biologists thought that proteins were the only biological molecules complex enough to carry genetic information, and that DNA provided only structural support.

In 1944, however, Rockefeller Institute researchers Oswald Avery, Colin McLeod, and Maclyn McCarty showed experimentally that DNA rather than protein was the carrier of hereditary information. In 1949, Austrian chemist Erwin Chargaff showed that the composition of DNA is not constant but varies from species to species—though the amount of A is always equal to the amount of T, and the amount of C is always equal to the amount of G. By 1950, it was generally agreed that Mendel's genes consisted of DNA, and Chargaff's ratios became an important clue for Watson and Crick in their efforts to unravel its structure.

Each of the four subunits of DNA consists of an identical sugar-phosphate group, to which is attached one of four different bases (A, T, C, or G). Relying in part on X-ray data from Rosalind Franklin, Watson and Crick proposed a double helix with the sugar-phosphate backbones on the outside and the bases on the inside. Using cardboard cutouts to represent the latter, Watson and Crick discovered that A on one strand would naturally pair with T on the other, and C on one strand would naturally pair with G on the other. The result would be two complementary strands: if the sequence on one were T-T-G-T, the sequence on the other would be A-A-C-A. This accounted for Chargaff's ratios. (Figure 6) The bonds holding each A-T pair and each C-G pair were relatively weak, so the two strands could be pulled apart and each could then serve as a template to synthesize another complementary one, resulting in two double helices with identical sequences. This suggested how genes could be copied for transmission to future generations.[3]

DNA: the Message

Genes are commonly regarded as DNA sequences that contain the information necessary to specify the sequences of the proteins needed by a living cell. Philosopher of science Stephen C. Meyer points out that DNA has three important properties that

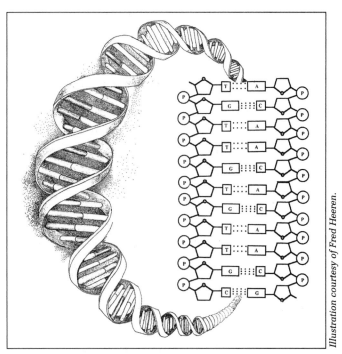

FIGURE 6. The Structure of DNA

Illustration courtesy of Fred Heeren.

enable it to carry this information. First, the subunits of DNA are like a four-letter alphabet. According to Meyer, they carry information "just like meaningful English sentences or functional lines of code in computer software." As Microsoft chairman Bill Gates once wrote: "DNA is like a computer program but far, far more advanced than any software ever created."[4]

Second, a typical gene is several hundred bases long, so its precise sequence is highly improbable. If this is true for a single gene, it is even truer for an entire organism. Estimates of the smallest number of genes needed to make a living cell start at 250.[5] If a minimal cell needs 250 genes, each several hundred bases long, then its gene sequences are so complex that the universe isn't old enough for them to have plausibly originated by chance. Therefore, the information in DNA is extremely complex.

Third, the information in the DNA is specified. A living cell needs not just any DNA, but DNA that encodes functional proteins. To be functional, a protein must have a very specific sequence. Francis Crick wrote in 1958: "For any particular protein the amino acids must be joined up in the right order." So proteins—and thus the DNA that encodes them—are both complex and specified. "By information," Crick wrote, "I mean the specification of the amino acid sequence of the protein ... [And since] in even a small bacterial cell there are probably a thousand different kinds of protein, each containing some hundreds of amino acids in its own rigidly determined sequence, the amount of hereditary information" in that cell "is quite considerable."[6]

So it is not the atoms and molecules in DNA that matter, but the fact that they are arranged in such a way that they can carry information. The secret of life is information.

But where does it come from?

The Origin of Biological Information

As a young geophysicist in the 1980s, Meyer became intrigued with a radical new idea that appeared in the epilogue to *The Mystery of Life's Origin*, a book by Charles B. Thaxton, Walter L. Bradley, and Roger L. Olsen. To explain the informational content of DNA, the authors adopted the principle of uniformity espoused by Darwin's contemporary, geologist Charles Lyell. They wrote: "By the principle of uniformity is meant that the kinds of causes we observe producing certain effects today can be counted on to have produced similar effects in the past." We observe in the present that intelligent agents can and do generate new information. "May not the principle of uniformity then be used," the authors asked, "to suggest that DNA had an intelligent cause at the beginning?"[7]

What Bill Gates said

"DNA is like a computer program but far, far more advanced than any software ever created."

—*The Road Ahead*, 1995

To investigate this idea further, Meyer left his job as a geophysicist and went to England to pursue a Ph.D. in the history and philosophy of science at Cambridge University. He eventually concluded from his studies that Thaxton, Bradley, and Olsen's intuition was correct.

Meyer observed that scientists explain past events by using an inference to the best explanation (Chapter Eight). Historical science typically relies on a uniformitarian appeal to causes that can be observed in the present. For example, a layer of ash in rock strata is better explained by a past volcanic eruption than by an earthquake, because we observe that the former produces ash layers in the present, while the latter does not. Following this line of reasoning, Meyer formulated a scientific inference to the best explanation for the origin of information in DNA. "We know from experience," he wrote, "that conscious intelligent agents can create informational sequences and systems." Since "we know that intelligent agents do produce large amounts of information, and since all known natural processes do not (or cannot), we can infer design as the best explanation of the origin of information in the cell."[8]

Origin-of-life research supports this inference. Although scientists have shown that some of the molecular building blocks of DNA, RNA, and protein can form under natural conditions, in the absence of living cells and without the aid of intelligent design, those building blocks do not spontaneously assemble into large information-carrying molecules. Sequences that carry biological information are always either purposefully synthesized by scientists or copied from existing sequences in living cells. Since the only cause known to be capable of producing such sequences in the present is intelligent design, it is reasonable for historical scientists to infer that an intelligence acted somehow in the past to produce the existing information-rich sequences in living cells.

Darwinists have raised several objections to Meyer's argument. First, they accuse him of arguing from ignorance, as though he infers design only when he doesn't know the cause of something. On the contrary,

Meyer explains: "Inferences to the best explanation do not assert the adequacy of one causal explanation merely on the basis of the inadequacy of some other causal explanation. Instead, they compare the explanatory power of many competing hypotheses to determine which hypothesis would, if true, provide the best explanation for some set of relevant data." Meyer continues, "it is not correct to say that we do not know how information arises." We know that it arises from intelligent agents.[9]

Second, some Darwinists have argued that DNA sequences result simply from the action of natural laws. But Meyer points out that the sequence of bases in DNA is not predetermined by the laws of physics and chemistry. As shown in Figure 6, if one sugar-phosphate group has a G attached to it, the next sugar-phosphate group could have a G, C, T or A. Although the sequence of bases on one strand determines the sequence of bases on the second strand, no physical or chemical laws dictate what the sequence will be in an isolated single strand of DNA.

An Article You're Not Supposed to Read

"The origin of biological information and the higher taxonomic categories," by Stephen C. Meyer, *Proceedings of the Biological Society of Washington* 117 (2004): 213–39. Available online at: http://www.discovery.org/scripts/viewDB/index.php?command=view&id=2177.

Indeed, if chemical affinities between the bases in one strand of DNA determined their sequence, the strand could not carry hereditary information. Consider what would happen if every time A occurred in a growing DNA sequence, it would drag (say) a T along with it, and every time C occurred it would be followed by (say) a G. The result would be the sort of repetitive sequence that biochemists before 1944 thought rendered DNA unfit as the carrier of genes. Chemist Michael Polanyi once wrote that if the precise sequence of bases in a DNA molecule were determined by chemical bonds between them, "then such a DNA molecule would have no information content...[The sequence] must be as physically indeterminate as the sequence of words is on a printed page."[10]

Meyer adds: "The information contained in an English sentence or computer software does not derive from the chemistry of the ink or the physics of magnetism, but from a source extrinsic to physics and chemistry altogether. Indeed, in both cases, the message transcends the properties of the medium. The information in DNA also transcends the properties of its material medium." And like the information in a sentence or a computer program, the information in DNA points to intelligence, because intelligent agency is "the only cause known to be capable of creating an information-rich system, including the coding regions of DNA, functional proteins, and the cell as a whole."[11]

Third, some Darwinists propose a scenario for the origin of life that, they say, could explain the origin of information-rich molecules without design. According to the "RNA world" scenario, a non-living mixture of relatively simple proteins and RNA molecules might reach a point at which the mixture could self-replicate. Natural selection could then act on the mixture to refine it and allegedly generate new information to produce a living cell.

Defending the RNA world scenario, Brown University biologist Kenneth R. Miller accuses Meyer of "intellectual desperation" for "failing to tell his readers of experiments showing that very simple RNA sequences can...self-replicate." Meyer responds that it is Miller who fails to tell people the whole story about the RNA world scenario. Synthesizing RNA under pre-life conditions has proven either difficult or impossible, and the RNA molecules Miller describes already contain complex specified information, the origin of which remains unexplained. Furthermore, even with intelligently designed molecules in a carefully controlled laboratory situation, RNA world research has not produced anything approaching the specified complexity in a living cell.[12]

Peer Review

When an article is submitted to a science journal, the editor may choose to accept or reject it outright. Watson and Crick's now-famous 1953 article on the structure of DNA was accepted by the editor of *Nature* without further review. More commonly, a journal editor sends an article to several outside referees ("peer reviewers") who advise the editor whether the article should be accepted as written, accepted only after revision, or rejected.

In 2003, Meyer submitted an article titled "The Origin of Biological Information and the Higher Taxonomic Categories" to the peer-reviewed *Proceedings of the Biological Society of Washington*. The article provided extensive references from the scientific literature to support Meyer's argument that DNA carries complex specified information that cannot be produced solely by natural processes such as mutation and selection. Relying on an inference to the best explanation, Meyer concluded that intelligent design was the cause of the enormous increase in biological information required to produce the major animal body plans in the Cambrian explosion.

Meyer wrote: "Analysis of the problem of the origin of biological information . . . exposes a deficiency in the causal powers of natural selection that corresponds precisely to powers that agents are uniquely known to possess. Intelligent agents have foresight. Such agents can select functional goals before they exist." Intelligent design theorists "are not positing an arbitrary explanatory element unmotivated by a consideration of the evidence. Instead, they are positing an entity possessing precisely the attributes and causal powers that the phenomenon in question requires."[13]

The editor of the *Proceedings of the Biological Society of Washington*, Richard M.

Lacking in intelligence

"What natural selection lacks, intelligent selection—purposive or goal-directed design—provides."

—**Stephen C. Meyer**, *Proceedings of the Biological Society of Washington*, 2004

von Sternberg, was a research associate at the Smithsonian Institution's National Museum of Natural History (NMNH) with two doctoral degrees in evolutionary biology. Following standard procedure, Sternberg sent Meyer's article to three reviewers, all of them evolutionary and molecular biologists at well-known institutions. The reviewers recommended that the article be published, though only after substantial revisions. Meyer revised his article in accordance with their recommendations, and the journal published it in August 2004.[14]

In recent years, a few ID-friendly articles have been published in peer-reviewed science journals.[15] Before 2004, none of those articles had explicitly defended intelligent design, and nobody had made much of a fuss over them. But when the *Proceedings of the Biological Society of Washington* published Meyer's article proposing intelligent design as an explanation for the origin of biological information, all hell broke loose.

Catch-23

Science journals regularly publish articles *attacking* intelligent design, but they routinely reject articles *defending* intelligent design. For example, Darwinists have criticized Michael Behe's arguments for ID (Chapter Ten) in many peer-reviewed science journals, including *Nature, Trends in Ecology and Evolution,* and the *Quarterly Review of Biology.* But those journals routinely refuse to publish Behe's responses. One journal editor to whom Behe submitted a response cited a reviewer who wrote: "In this referee's judgment, the manuscript of Michael Behe does not contribute anything useful to evolutionary science." The editor of another journal wrote to Behe: "As you no doubt know, our journal has supported and demonstrated a strong evolutionary position from the very beginning, and believes that evolutionary explanations of all structures and phenomena of life are possible and inevitable. Hence a position such as yours ... cannot be appropriate for our pages."[16]

In Joseph Heller's classic novel about World War II, *Catch-22,* an aviator could be excused from combat duty for being crazy. But a rule specified that he first had to request an excuse, and anyone who requested an excuse from combat duty was obviously not crazy, so such requests were invariably denied. The rule that made it impossible to be excused from combat duty was called "Catch-22."

Darwinists use a similar rule—I call it "Catch-23"—to exclude intelligent design from science: intelligent design is not scientific, so it can't be published in peer-reviewed scientific journals. How do we know it's not scientific? Because it isn't published in peer-reviewed scientific journals.[17] Catch-23!

The 2004 publication of Meyer's article shattered the rule. It also alarmed Darwinists at the Smithsonian Institution (SI), with which the Biological Society of Washington (BSW) is loosely affiliated. Smithson-

Frankenstein's castle?

To Dr. Richard Sternberg, concerning his treatment by the Smithsonian Institution (SI) after publishing a peer-reviewed article by Dr. Stephen C. Meyer on intelligent design: "Our preliminary investigation indicates that retaliation came in many forms. It came in the form of attempts to change your working conditions.... During the process you were personally investigated and your professional competence was attacked. Misinformation was disseminated throughout the SI and to outside sources. The allegations against you were later determined to be false. It is also clear that a hostile work environment was created with the ultimate goal of forcing you out of the SI."

—U.S. Office of Special Counsel, 2005

ian Darwinists teamed up with the militantly pro-Darwin National Center for Science Education (NCSE) to control the damage to their cause. NCSE staffers sent long, detailed e-mails attacking Meyer's article to high officials at the Smithsonian. The NCSE then worked closely with Smithsonian employees to develop a strategy of character assassination to punish Sternberg for publishing the article. To protect himself, Sternberg lodged a complaint with the U.S. Office of Special Counsel (OSC), established by Congress to investigate such cases.[18]

In August 2005, the OSC sent Sternberg a letter notifying him that a recent administrative decision had removed his case from their jurisdiction, but confirming that "members of NCSE worked closely with SI and NMNH members in outlining a strategy to have you investigated and discredited," noting that "OSC questions the use of appropriated funds to work with an outside advocacy group for this purpose." The OSC letter also confirmed that the management of the Smithsonian had falsely accused Sternberg of mishandling specimens in his research and of violating *Proceedings of the Biological Society of Washington* policies in the publication of Meyer's article. These accusations "were published to several outside organizations," severely damaging Sternberg's reputation. The managers later admitted that the accusations were false, but the OSC saw no evidence that "any effort was made to recall or correct these comments once the truth was known." There were other abuses, too, but since the OSC lost jurisdiction over the Sternberg case "the SI is now refusing to cooperate with our investigation." Nevertheless, the OSC concluded that the management of the publicly funded Smithsonian Institution had deliberately "created a hostile working environment" for Sternberg, hoping that he would "leave or resign."[19]

To investigate the Darwinists' accusation that Sternberg had circumvented the normal peer-review process, the president of the Council of the BSW reviewed the file, and he found that the peer review had been properly conducted. Nevertheless, the council subsequently issued a statement declaring that "the Meyer paper does not meet the scientific standards of the *Proceedings*."[20] Although the BSW stopped short of formally retracting the article, the Darwinists did not end their ruthless campaign of character assassination against Sternberg.

Catch-23 is still enforced by most science journals, but it is now supplemented with an additional rule: if intelligent design theorists *do* manage to publish in a peer-reviewed science journal, Darwinists will make sure the editor suffers grievously for it.

Chapter Ten

DARWIN'S BLACK BOX

On a midsummer day in 1906, on an island in one of Minnesota's ten thousand lakes, Norwegian immigrant and self-taught engineer Ole Evinrude was having a picnic with his fiancée, Bess. After he and his beloved finished their lunch, Ole rowed several miles in ninety-degree heat to get her some ice cream. Although he was strong and energetic, Ole decided then and there that it was time to put a motor on his boat. He proceeded to invent the first commercially successful outboard motor, and in 1911 he secured a patent for what he called his Marine Propulsion System.[1]

What Evinrude didn't know was that the water under his boat teemed with microscopic creatures that already had outboard motors far superior to his. Many bacteria have long whip-like appendages called flagella, which rotate at speeds up to 100,000 rpm and function as tiny marine propulsion systems. Using Evinrude's putt-putt, a small rowboat could travel ten miles per hour, or about one and a half boat lengths per second. Using flagella, the common intestinal bacterium *E. coli* moves about ten body lengths per second—more than six times as fast (to scale) as Evinrude's boat.

The motor that drives the flagellum is so small that biologists were unable to study it until electron microscopes were invented in the 1930s. We now know that the motor of the bacterial flagellum is an intricate

Guess what?

- Anything that natural selection cannot make in small successive steps is a problem for Darwinism, and biochemist Michael Behe has identified such things in living cells.
- Behe argues that such "irreducibly complex" features point to intelligent design.
- Darwinists claim that Behe's argument is not scientific because it is untestable; and besides, they say, they have tested it and proven it wrong.

molecular machine. After comparing the flagellar motor to various other molecular motors in living cells, Brandeis University biologist David DeRosier wrote in 1993: "More so than other motors, the flagellum resembles a machine designed by a human."[2]

So the motor of the bacterial flagellum *looks* designed. But is it *actually* designed?

Irreducible Complexity

In a 1986 scientific monograph, theoretical biologist Michael J. Katz of Case Western Reserve University wrote: "Contemporary organisms are quite complex, they have a special and an intricate organization that would not occur spontaneously by chance. The 'universal laws' governing the assembly of biological materials are insufficient to explain our companion organisms: one cannot stir together the appropriate raw materials and self-assemble a mouse." According to Katz: "There are useful scientific explanations for these complex systems, but the final patterns that they produce are so heterogeneous that they cannot effectively be reduced to smaller or less intricate predecessor components." Katz concluded: "These patterns are, in a fundamental sense, irreducibly complex."[3]

Katz regarded irreducible complexity as a problem for evolutionary biology, but ten years later Lehigh University biochemist Michael J. Behe (independently of Katz's work) took things a step further. In his 1996 book *Darwin's Black Box*, Behe (pronounced bee-hee) argued that irreducible complexity not only refutes Darwin's theory, but also provides evidence for intelligent design.

In *The Origin of Species*, Darwin wrote: "If it could be demonstrated that any complex organ existed which could not possibly have

A Book You're Not Supposed to Read

Darwin's Black Box: The Biochemical Challenge to Evolution, Tenth Anniversary Edition, by Michael J. Behe; New York: The Free Press, 2006.

been formed by numerous, successive, slight modifications, my theory would absolutely break down." In *Darwin's Black Box*, Behe wrote: "What type of biological system could not be formed by 'numerous successive, slight modifications? Well, for starters, a system that is irreducibly complex. By irreducibly complex I mean a single system composed of several well-matched interacting parts that contribute to the basic function, wherein the removal of any one of the parts causes the system to effectively cease functioning."[4]

To illustrate what he meant by "irreducibly complex," Behe used a mousetrap. A common household mousetrap consists of at least five parts: a flat wooden base on which the other parts are mounted, a metal hammer to crush the mouse, a wire spring connected to the base and the hammer, a pressure-sensitive catch on which the bait is placed, and a metal bar connected to the catch, which holds the hammer back when the trap is charged. If any one of these components (the base, hammer, spring, catch, or holding bar) is removed, then the mousetrap does not catch mice. Because it is necessarily composed of several parts, the mousetrap is irreducibly complex.

According to Behe, the irreducible complexity we see in the mousetrap is a hallmark of design. Even though some or all of its parts—such as the spring or the wooden base—could serve other purposes, they would not be arranged as they are in the mousetrap except for the purpose of catching mice. And even though a mousetrap could be constructed in a series of successive modifications, as Behe's critic John H. McDonald has shown, those modifications would have to be guided by an intelligent agent. Darwin's writings make it clear that by "successive, slight modifications" he meant *unguided* modifications, so McDonald's example (as he acknowledges) cannot serve as an analogy for Darwinian evolution.[5]

In any case, a mousetrap is not a living thing. Do living things contain irreducibly complex features? If they do, then they pose a challenge to Darwinian evolution. Natural selection cannot assemble parts for the purpose of producing future functions; it can only preserve features that already have functions. So a feature that is irreducibly complex—that does not function until all its parts are in place—cannot be assembled by natural selection. In Behe's view, such a feature points to intelligent design.

Behe proceeded to describe a number of features in living cells that are irreducibly complex. Three such features are the light-sensing mechanism in our eyes, the human blood-clotting system, and the microscopic motor that propelled the bacteria in the lake under Ole Evinrude's boat.

Seeing and Clotting

When a machine or structure or process performs a function, but its inner workings are unknown, scientists call it a "black box." To Darwin, the cell was a black box. He and his contemporaries were almost completely ignorant of the inner workings of the cell; they believed that a cell was simply a blob of jelly with special properties. Modern microbiologists and biochemists, however, have discovered in living cells an amazing world of highly efficient microscopic machines and precisely integrated biochemical systems.

In *The Origin of Species,* Darwin acknowledged that the human eye presented a problem for his theory. How could such a complex organ have evolved through successive slight modifications? Although he did not know what pathway evolution took in constructing the eye, Darwin pointed to a variety of animals that had eyes ranging from a simple light-sensitive spot to the complex vertebrate camera eye, and he speculated that similar organs might have been intermediates in the evolution of the human eye.

For Darwin, the "light-sensitive spot" was a black box. Only after the advent of modern biochemistry did scientists discover how complex that spot really is.

When light strikes the human retina it is absorbed by a molecule that alters an attached protein, which then initiates what biochemists call a "cascade"—a precisely integrated series of molecular reactions—that in this case causes a nerve impulse to be transmitted to the brain. If any molecule in the cascade is missing or defective, no nerve impulse is transmitted; the person is blind. Since the light-sensing mechanism doesn't function at all unless every part is present, it is irreducibly complex. The fossil record cannot tell us, and no evolutionary biologist has explained, how all these molecules could have assembled themselves to produce the light-sensitive spot that was the starting point for Darwin's speculation.

In 2005, University of Chicago evolutionary biologist Jerry A. Coyne wrote that Behe's argument has a fatal flaw: "Darwin brilliantly addressed this argument by surveying existing species to see if one could find functional but less complex eyes that not only were useful, but also could be strung together into a hypothetical sequence showing how a camera eye might evolve. If this could be done—and it can—then the argument for irreducible complexity vanishes."[6]

But Coyne's comments had nothing to do with Behe's argument, which is based on biochemistry rather than anatomy. "It is no longer enough," Behe wrote in 1996, "to consider only the anatomical structure of whole eyes, as Darwin did in the nineteenth century (and as popularizers of evolution continue to do today). . . . Anatomy is, quite simply, irrelevant to the question of whether evolution could take place on the molecular level." Coyne was merely repeating Darwin's old argument, as though modern biochemistry did not exist.[7]

Another example of irreducible complexity in living cells is the human blood-clotting cascade. If a person suffers even a minor cut, it is essential that a clot form to stop the bleeding and close the wound; otherwise, the person might bleed to death. Yet if a clot were to form in our blood vessels when it's not needed, that could cause death, too. The clot itself is not all that complicated, but the blood-clotting cascade consists of more than a dozen protein molecules that must interact sequentially with each other to produce a clot *only* at the right time and place. Each protein is extremely complex in its own right, but it is the cascade that Behe identified as irreducibly complex, because all of the molecules must be present for the system to work. If even one is missing (as in the case of hemophilia), the system fails.

University of California–San Diego biochemist Russell Doolittle, an expert on blood clotting, claimed in 1997 that experiments had proven Behe wrong. The experiments showed that if one component of the cascade is knocked out in one group of mice and another component is knocked out in another group, both groups lack functional clotting systems; but (Doolittle claimed) "when these two lines of mice were crossed ... [then] for all practical purposes, the mice lacking both genes were normal!" He concluded: "Contrary to claims about irreducible complexity, the entire ensemble of proteins is *not* needed."[8]

But Doolittle had misread the scientific articles on which he based his argument. When mice from the two abnormal groups were crossed, their offspring were *not* normal, but lacked a functional clotting system and

Behe ever so humble

"The conclusion of intelligent design flows naturally from the data itself—not from sacred books or sectarian beliefs. Inferring that biochemical systems were designed by an intelligent agent is a humdrum process that requires no new principles of logic or science. It comes simply from the hard work that biochemistry has done over the past forty years, combined with consideration of the way in which we reach conclusions of design every day."

—**Michael J. Behe,** *Darwin's Black Box,* 2006

suffered from frequent hemorrhages. The fact that an expert such as Doolittle could come up with nothing better than experiments that actually *support* Behe's argument convinced Behe "that there are indeed no detailed explanations for the evolution of blood clotting in the literature and that, despite Darwinian protestations, the irreducible complexity of the system is a significant problem for Darwinism."[9]

The eye's light-sensing mechanism and the human blood-clotting cascade are only two of several biochemical systems that Behe cites to support his argument. But the most famous example of irreducible complexity in living cells—some have called it the poster child of intelligent design—is mechanical rather than biochemical.

The Bacterial Flagellum

The bacterial flagellum is a long, hair-like filament. The common intestinal bacterium *E. coli* has on average six to twelve extending from its body. When the flagella turn in one direction they bundle together to form a long, rapidly rotating whip that propels the organism through the surrounding liquid; when the flagella reverse direction the whip unravels and the organism stops abruptly and tumbles. At the base of each flagellum is a proton-driven motor that can turn thousands of times a minute and reverse direction in a quarter turn. The motor's drive shaft is attached to a rotor that turns within a stator, and the entire assembly is anchored in the cell wall by various bushings. The filament itself is attached to the drive shaft by a hook that functions as a universal joint so the flagellum can twist as it turns. (Figure 7) The operation of the flagellum is linked to a system that enables the organism to sense and follow chemicals in its surroundings.

By knocking out genes and screening for cells that can no longer move, researchers have identified all the gene products (proteins) required for assembly and operation of the flagellum. Remove any of them, and it

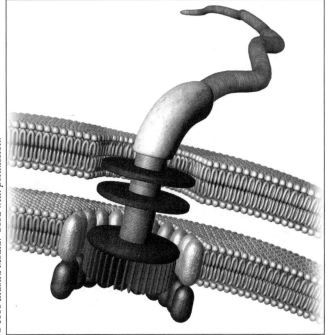

FIGURE 7.
The Motor of the
Bacterial Flagellum

stops working. Thus the motor of the bacterial flagellum meets Behe's criterion for irreducible complexity.[10]

Brown University biologist Kenneth R. Miller has attempted to refute Behe by arguing that the motor of the bacterial flagellum is not irreducibly complex. Bacterial cells form their long flagellar filaments by secreting protein subunits through the hollow drive shaft of the flagellar motor. A few years ago, microbiologists discovered that some pathogenic bacteria use a similar hollow shaft as a needle to inject toxins into the cells of their victims. The similar (though not identical) apparatus in each case is called the type III secretory system, or TTSS.

Miller argues that because the TTSS resembles part of the flagellum but has a different biological function, the flagellum itself is not irreducibly complex. Furthermore, he argues, the pre-existing TTSS was probably "co-opted" by Darwinian evolution to produce the flagellum. Miller concludes: "What this means is that the argument for intelligent design of the flagellum has failed."[11]

In *Darwin's Black Box*, however, Behe had already acknowledged that irreducibly complex systems sometimes contain subsets that perform other functions in other contexts. For example, a mechanic could remove the gasoline engine from an outboard motor and run it by itself, but the outboard motor can't function without it. Removing the engine doesn't refute the irreducible complexity of the outboard motor; in fact, it confirms it. In a 2004 response to critics, Behe wrote that Miller was "switching the focus from the function of the system to act as a rotary propulsion machine to the ability of a subset of the system to transport proteins across a membrane. However, taking away the parts of the flagellum certainly destroys the ability of the system to act as a rotary propulsion machine, as I have argued. Thus, contra Miller, the flagellum is indeed irreducibly complex."[12]

As for Miller's claim that the TTSS was probably co-opted by Darwinian evolution to produce the flagellum—University of Idaho microbiologist Scott Minnich disagrees. Minnich is an expert on plague bacteria (which use the TTSS to attack their victims), and in 2004 he spent five months as a volunteer in Iraq with an elite team searching for biological weapons. While in Iraq, he and Stephen C. Meyer contributed a paper to an international biology conference in Greece in which they pointed out that the flagellar motor consists of several dozen proteins that are not present in the TTSS but are "unique to the motor and are not found in any other living system." They asked: "From whence, then, were these proteins co-opted?"[13]

Another problem for Miller's co-option argument is that flagella probably existed before the TTSS. Several lines of evidence support this conclusion. Minnich and Meyer conclude: "If anything, the pump evolved from the motor, not the motor from the pump."[14]

Which Moscow Is This?

In a pattern that is becoming increasingly clear (and will become even clearer in Chapter Sixteen), both Behe and Minnich have come under fire on their home campuses for defending intelligent design. According to the chairman of Behe's department at Lehigh University in Pennsylvania, the faculty there "actively and forcefully" condemns Behe's work.[15] At the University of Idaho, where Minnich is a professor of microbiology, the heat has been even more intense.

A few months after Minnich returned home from Iraq to his family in Moscow, Idaho, a school district in Dover, Pennsylvania, decided to inform its high school biology students that there is a controversy over intelligent design and that the theory of evolution should be studied with an open mind. The American Civil Liberties Union (ACLU) took the school district to court, claiming that the statement was unconstitutional. (We will take a closer look at this case in Chapters Thirteen and Sixteen.) The Dover school district sought expert witnesses to defend its constitutional right to freedom of speech, and Minnich (among others) volunteered to help—on his own time.

In October 2005, just before Minnich was scheduled to testify in Pennsylvania, University of Idaho president Timothy P. White issued an edict prohibiting the teaching of "views that differ from evolution . . . in our life, earth, and physical science courses."[16] A week after President White issued his edict, the university hosted a seminar by Darwinist Eugenie Scott, titled "Why Scientists Reject Intelligent Design." Scientist Scott Minnich was not invited to participate.

Actually, Minnich had never taught his students that Darwinism is wrong or intelligent design is right. Quite reasonably, though, he expected the academic establishment to respect his freedom to encourage students to think critically about this subject—or at least to respond to Eugenie Scott in an open forum.

Apparently academic freedom doesn't extend to critics of Darwinism. When University of Colorado professor Ward Churchill called victims of the terrorist attacks on Sep-

Designer genes

"Molecular machines display a key signature or hallmark of design, namely, irreducible complexity. In all irreducibly complex systems in which the cause of the system is known by experience or observation, intelligent design or engineering played a role in the origin of the system. . . . We find such systems within living organisms."

Scott A. Minnich and **Stephen C. Meyer**, Second International Conference on Design & Nature, 2004

tember 11, 2001, "little Eichmanns," the American Association of University Professors (AAUP) defended *his* academic freedom, reaffirming the AAUP's commitment "to preserving and advancing principles of academic freedom in this nation's colleges and universities. Freedom of faculty members to express views, however unpopular or distasteful, is an essential condition of an institution of higher learning that is truly free." But when word of President White's edict reached Jonathan Knight, director of the AAUP's Office of Academic Freedom, Knight said: "Academic freedom is not a license to teach anything you like."[17] In the Orwellian thinking of the AAUP, all unpopular views are equal, but some are more equal than others.

Defenders of President White's edict pointed to a "consensus" of scientists that intelligent design is wrong. But how could there be a "consensus" if qualified scientists such as Minnich and Behe are excluded from voting? This sounds suspiciously like those "unanimous" elections for which the former Soviet Union became notorious. Just as truth could not be decided by the Communist Party in Moscow, Russia, so it cannot be decided by the Darwinist Party in Moscow, Idaho.

Chapter Eleven

✵✵✵✵✵✵✵✵✵✵✵✵✵✵

WHAT A WONDERFUL WORLD

Neem ka Thana is a small town about one hundred miles from Delhi, India. Accessible by train but barely reachable by car, the village is not a big tourist attraction. At sunrise on October 24, 1995, however, Neem ka Thana was filled with foreign visitors. They were astronomers, loaded with cameras and scientific instruments, who had come from all over the world to study a total solar eclipse.[1]

One of them was Guillermo Gonzalez, a young Cuban immigrant to the United States who had recently earned a Ph.D. in astrophysics at the University of Washington. Gonzalez was in Neem ka Thana at the invitation of the Indian Institute of Astrophysics to photograph the eclipse and measure changes in atmospheric conditions as the Earth passed through the moon's shadow. He later wrote: "To experience a total solar eclipse is much more than simply to see it. The event summons all the senses. The dramatic drop in temperature was just as much a part of it as the blocked sun and the 'oohs' and 'aahs' from the crowd."[2]

What brought astronomers such as Gonzalez from all over the world to watch an eclipse that lasted less than a minute? In a total solar eclipse, the moon exactly covers the face of the sun, leaving only its fiery outer atmosphere visible from Earth. Studying that outer atmosphere enables astronomers to discover much of what they now know about the stars. If the moon were smaller or larger, or closer or farther

Guess what?

- The Earth's place in the universe is so remarkably well suited for life and scientific discovery that it suggests design.
- Opponents of design claim that our universe is just the luckiest of an infinite number of universes—for which there is no evidence.
- When Iowa State University astronomer Guillermo Gonzalez wrote about cosmic evidence for intelligent design, a militant atheist in the religious studies department organized a campaign against him.

119

away, such discoveries would have been delayed, perhaps indefinitely. It's as though the size and orbit of the moon were tailor-made for science.

As an astronomer, Gonzalez already knew that many factors contributed to making Earth the best place in the solar system for life. Now he realized that Earth was also the best place in the solar system to make scientific discoveries about the universe. It seemed unlikely that the conjunction of the two was due to chance. To Gonzalez, it was evidence of intelligent design.

Cosmic Fine-Tuning

The fundamental constants of the universe seem remarkably fine-tuned for life. For example, if the strength of gravity were weaker by only one part in a trillion trillion trillion, the universe would have expanded so quickly after the Big Bang that no galaxies or planets would have formed. On the other hand, if gravity were *stronger* by only one part in a trillion

In the end is my beginning

"The total number and diversity of possible chemical structures that may be constructed out of carbon, oxygen, hydrogen, and nitrogen is virtually unlimited. Almost any imaginable chemical shape and chemical property can be derived. Together these elements form what is in effect a universal chemical constructor kit.... It is as if from the very moment of creation the biochemistry of life was already preordained in the atom-building process, as if Nature were biased to this end from the beginning."

—Michael J. Denton, *Nature's Destiny*, 1998

trillion trillion, the universe would have quickly collapsed back on itself. Either way, we would not exist.

Other finely tuned fundamental constants include the strength of the electromagnetic force, the ratio of the masses of protons and electrons, and the strong nuclear force that holds atomic nuclei together. In fact, there are over a dozen constants that must have precisely the values they have in order to make the universe habitable. Oxford physicist Roger Penrose has calculated that the odds against all these constants having just the right values are one followed by trillions and trillions more zeroes than there are elementary particles in the universe. It would be impossible even to write out such an enormous number.[3]

The chemical elements also seem uniquely suited for life. For example, carbon is unusual in its ability to combine chemically not only with itself but also with many other elements, making possible the vast number of complex compounds needed by living cells. Several other elements—most notably hydrogen, oxygen, nitrogen, and phosphorus—are uniquely suited to combining with carbon to form biologically active molecules. As molecular biologist Michael Denton wrote in 1998: "It is as if from the very moment of creation the biochemistry of life was already preordained in the atom-building process."[4]

Hydrogen and oxygen also combine to form water. Most of the chemical reactions necessary for life can take place only in liquid water. Water's ability to absorb and retain heat also buffers living things from sudden temperature changes. According to Denton: "Water is uniquely and ideally adapted to serve as the fluid medium for life on Earth in not just one, or many, but in *every single one* of its known physical and chemical characteristics."[5]

The properties of the elements follow from the universal constants, so cosmic fine-tuning results not only in a habitable universe, but also in elements uniquely suited for life. Some people have attempted to explain these remarkable coincidences by invoking "the anthropic principle." We should not be surprised by fine-tuning, they say, because if the universe were not fine-tuned for life we would not be here to observe it. This hardly counts as an explanation. Philosopher John Leslie uses the metaphor of a condemned man who survives a firing squad composed of a thousand expert marksmen. According to Leslie, saying that the anthropic principle explains cosmic fine-tuning would be like the condemned man saying: "If the thousand men of the firing squad hadn't all missed me then I shouldn't be here to discuss the fact, so I've no reason to find it curious."[6]

A second conceivable explanation is that the universal constants have their precise values because of natural law or necessity. But the constants are essential parts of natural laws; they do not follow from those laws. Laws describe regularities, and since we know of only one universe we cannot observe regularities in whatever processes form universes.

A third explanation is that there must be an infinite number of parallel universes with fundamental constants having all possible values, and ours just happens to be the lucky one. The problem with this "multiverse" explanation is that we have no evidence—nor can we—of other universes.

Other scientists and philosophers maintain that cosmic fine-tuning points to intelligent design. By employing an inference to the best explanation (Chapter Eight), we can see that cosmic fine-tuning is much more likely if the universe is designed than if it is not.[7]

Our Privileged Planet

Cosmic fine-tuning is remarkable, but it is just the beginning. Even in a finely tuned universe, most places are uninhabitable. It turns out that the Earth's location in the cosmos is unusually well-suited for life.

First, most galaxies could not support life. In elliptical galaxies, stars have irregular orbits that take them through regions of the galaxy that are hostile to life. In spiral galaxies such as our Milky Way, stars follow more orderly orbits.

Second, our place in the Milky Way is just right. If we were closer to the galactic center we would be bombarded by dangerous radiation and comets. If we were closer to the edge, the sun would have fewer of the heavy elements needed to form planets. As it is, we are in a relatively narrow "Galactic Habitable Zone."[8]

Third, our sun and solar system are well suited to life. The Earth's distance from the sun, and its near-circular stable orbit, provide just the right conditions for liquid water, which is necessary for life. Just as the solar system is in the Milky Way's Galactic Habitable Zone, so Earth is in the solar system's "Circumstellar Habitable Zone."

Fourth, the moon stabilizes the tilt of the Earth's axis, preventing wild fluctuations in temperature. The moon (together with the sun) also contributes to the Earth's tides, which mix nutrients from the land with the oceans and help to drive ocean currents that regulate our climate.

Finally, there is the Earth itself. If our planet were much smaller, it would not be able to hold an atmosphere or generate the strong magnetic field that protects us from cosmic rays. If the Earth were much larger, the increased gravity would mean more atmospheric pressure and less surface relief; in fact, the planet might be entirely covered in water.[9]

Not only is Earth especially suited for life, but it is also well situated for scientific discovery. Because the Milky Way is a spiral galaxy, it is relatively flat, so we can observe distant galaxies that would otherwise be obscured by dust and stars in our own galaxy. And Earth's position in the Milky Way, about halfway between the galactic center and its visible edge, is just about

A Book You're Not Supposed to Read

The Privileged Planet: How Our Place in the Cosmos Is Designed for Discovery, by Guillermo Gonzalez and Jay W. Richards; Washington, DC: Regnery Publishing, 2004.

ideal for making astronomical observations, giving us a fairly clear view of nearby stars as well as distant galaxies.

The characteristics of our solar system are also well suited to scientific discovery. By observing a star's location at one time of year and observing it again six months later, the shift in its position relative to more distant stars enables astronomers to calculate its distance. The relative sizes and distances of the Earth, moon, and sun are important factors in making such measurements. The same parameters also make total solar eclipses possible, which make it possible for us to discover important characteristics of the stars.

The most habitable places in the universe are also the best places to make scientific discoveries about the universe. According to Gonzalez and Jay W. Richards's *The Privileged Planet*: "There's no obvious reason to assume that the very same rare properties that allow for our existence would also provide the best overall setting to make discoveries about the world around us. We don't think this is merely coincidental. It cries out for another explanation, an explanation that suggests there's more to the cosmos than we have been willing to entertain or even imagine." They conclude that the correlation between the factors needed for complex life and the factors needed to do science "forms a meaningful pattern" that "points to purpose and intelligent design in the cosmos."[10]

Intelligent Design at the Smithsonian

Soon after Gonzalez and Richards published *The Privileged Planet*, Illustra Media produced a one-hour film based on the book. In October 2004, the film premiered to a standing-room-only crowd at Seattle's Museum of Flight.[11]

The Discovery Institute in Seattle (with which Gonzalez and Richards were both affiliated) then approached the National Museum of Natural

History (NMNH) at the Smithsonian Institution in Washington, D.C. to arrange a showing there. The NMNH—the same organization that persecuted Richard Sternberg for publishing an article about intelligent design—routinely makes its auditorium available to outside groups in exchange for a donation. Following standard procedure, NMNH staff reviewed the film to make sure it complied with the museum's policy excluding events of a religious nature. In April 2005, the NMNH agreed to co-sponsor a showing of *The Privileged Planet* on June 23, in exchange for a $16,000 donation from the Discovery Institute. In May, invitations were sent to several hundred people in North America announcing: "The director of the National Museum of Natural History and Discovery Institute cordially invite you to the national premiere and evening reception of *The Privileged Planet: The Search for Purpose in the Universe*."[12]

> ### Location, location, location
>
> "In many ways, ours is the optimal galaxy for life … [and] we're only now beginning to appreciate how much our solar system's configuration is not only rare but also surprisingly crucial for life."
>
> **—Guillermo Gonzalez** and **Jay W. Richards**

When Darwinists learned of the upcoming event, they went ballistic. Their Internet blogs, bristling with indignation, urged readers to send protests to the Smithsonian. The *New York Times* announced: "Smithsonian to Screen a Movie that Makes a Case Against Evolution." Yet the film (like the book) is compatible with cosmic evolution, and it says nothing about biological evolution.[13]

After being hammered by enraged Darwinists the Smithsonian threw in the towel. Lucy Dorrick, associate director for development and special events at the NMNH, wrote to the Discovery Institute: "Upon further review, the Museum has determined that the content of the film is not consistent with the mission of the Smithsonian Institution's scientific research." NMNH spokesman Randall Kremer told *The Scientist*: "The scientific content for the most part is accurate," but "the science is used to draw a philosophical conclusion."[14]

Discovery Institute's Jonathan Witt pointed out that the Smithsonian had had no problem sponsoring "Cosmos Revisited: A Series Presented in the Memory of Carl Sagan" in 1997. "Sagan's 'Cosmos' series," noted Witt, "is famous for its opening dictum, 'The Cosmos is all that is, or ever was, or ever will be.' Why didn't the Smithsonian have a problem promoting this little philosophical flourish?" Witt concluded: "The Smithsonian has been given over, lock, stock, and barrel, to Sagan's metaphysical vision for decades. The one difference now is that they're explicitly stating that not only do they privilege Sagan's materialist metaphysic, they will block any scientific argument that suggests a contrary conclusion."[15]

But the Smithsonian had already signed a contract for the event. "Due to this fact," wrote Dorrick, "we will, of course, honor the commitment made to provide space for the event to the Discovery Institute, but the museum will not participate or accept a donation for the event." The NMNH returned $11,000 of the $16,000 donation to the Discovery Institute, keeping $5,000 for expenses.

The event itself was a huge success. On June 23, 2005, a capacity crowd filled the National Museum of Natural History's Baird Auditorium. Afterwards, Gonzalez and Richards answered questions from scientists, students, journalists, and legislators, then hosted a reception in the museum's Hall of Geology, Gems, and Minerals.[16]

The controversy also had its lighter moments. When he first heard of it at the end of May, atheist James Randi offered to pay the NMNH $20,000 to cancel the event. Randi's clumsy attempt to bribe the Smithsonian to censor the film amused mathematician David Berlinski, a critic of Darwinism and fellow of the Discovery Institute living in Paris. Berlinski wrote to Randi and threatened, tongue in cheek, to show *The Privileged Planet* in Europe unless Randi also paid *him* $20,000. For "the right price" Berlinski said, the Discovery Institute would make sure the film "disappears itself, if you catch my drift. You get to keep the negatives, we keep the director's cut in our safe for insurance. Is this some sort of deal, or what?"[17]

Randi did not respond.

Is This Heaven?

In 2001, Guillermo Gonzalez had left his postdoctoral research position at the University of Washington to take a job as an assistant professor of astronomy and physics at Iowa State University. Things were going well there until the June 2005 screening of *The Privileged Planet* at the Smithsonian.

Soon after the Smithsonian event, Iowa State professor Hector Avalos circulated a faculty petition to "reject all attempts to represent intelligent design as a scientific endeavor." The petition, which was signed by nearly 120 Iowa State professors, criticized "advocates of intelligent design" for claiming that "the position of our planet and the complexity of particular life forms and processes are such that they may only be explained by the existence of a creator or designer of the universe." Within a few weeks, many faculty members at the University of Iowa and the University of Northern Iowa had also signed the petition.[18]

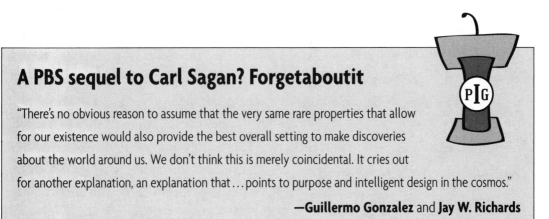

A PBS sequel to Carl Sagan? Forgetaboutit

"There's no obvious reason to assume that the very same rare properties that allow for our existence would also provide the best overall setting to make discoveries about the world around us. We don't think this is merely coincidental. It cries out for another explanation, an explanation that . . . points to purpose and intelligent design in the cosmos."

—**Guillermo Gonzalez** and **Jay W. Richards**

"We want to make sure the public and the university start to voice their opposition to intelligent design," Avalos told the *Iowa State Daily*. He also told the *Chronicle of Higher Education*: "We certainly don't want to give the impression to the public that intelligent design is what we do." And he was quoted by the Associated Press as saying: "A lot of people were concerned that Iowa State could become a place being marketed where intelligent design research was taking place and that it had some validity in school curricula."[19]

Although the petition did not mention Gonzalez by name, it was obviously aimed at him. Yet he has never taught intelligent design in his classes. He does assert that "it properly falls within science" because its methods are scientific and it does not start with religious assumptions, but he also considers it too new and controversial to teach without the support of his astronomy colleagues. Like Michael Behe and Scott Minnich, Gonzalez defends intelligent design only outside the classroom, on his own time. Yet he, like Behe and Minnich, is condemned by colleagues who in other situations boast about their commitment to academic freedom.[20]

Avalos accuses Gonzalez of having a hidden religious agenda. Others have accused him of academic fraud. "I didn't expect this level of vitriol, this level of intense hostility," Gonzalez told the *Des Moines Register*. At the University of Iowa, about a hundred miles east of Iowa State, physics professor Frederick Skiff agreed to discuss intelligent design in a campus forum with three Darwinists. Skiff was ridiculed and insulted, and he was given very little time to respond. "I

Science and Halloween

"No one has offered any evidence that ID has ever been taught in an ISU class by anyone. Gonzalez says he hasn't taught it. Can it really be that all these defenders of science are attacking something entirely imaginary? The lack of a witch has never deterred the witch hunters, but a witch hunt is a poor model for scientific inquiry."

—Iowa State sociology professor **Dave Schweingruber**, 2005 have never faced such blatant hostility and dishonesty at the hands of colleagues before," he said afterward.[21]

Someone unfamiliar with Iowa State might think that Avalos is a scientist—perhaps, like Gonzalez, a physicist or astronomer—concerned with defending the integrity of his discipline from attacks by religious fundamentalists. But Hector Avalos is a professor of religion! In this bizarre drama, it is the scientist who is arguing for design in the universe, and the religion professor who is trying to silence him. The irony was not lost on Iowa State sociology professor Dave Schweingruber, who wrote to a local newspaper: "What is Avalos' objection to Gonzalez's work? He told the *Des Moines Register* that he knows ID is religion and not science because 'I'm a Biblical scholar.' So Iowa State has one thing in common with unaccredited Bible colleges and medieval heresy tribunals: Our Bible scholars think they can tell our astronomers how to do their jobs." Schweingruber concluded: "A witch hunt is a poor model for scientific inquiry."[22]

But the irony doesn't stop there. Religion professor Hector Avalos is a militant atheist. He is the founder and faculty advisor of the Iowa State University Atheist and Agnostic Society, which "is intended to provide an educational and support system for students who believe that one can live a fulfilling, productive, and ethical life without religion."[23] So Iowa taxpayers are spending hundreds of millions of dollars every year on an institution that entrusts the teaching of religion to someone who tells students they're better off without it—and who thinks it's his job to tell an astronomer how to do science.

Is this heaven? No, it's Iowa.

Chapter Twelve

✧✧✧✧✧✧✧✧✧✧✧✧✧✧

IS ID SCIENCE?

I n their eagerness to oppose intelligent design, Darwinists try to exclude it from serious consideration by arguing that it is not science.

In 2004, American Society for Cell Biology president Harvey F. Lodish wrote that intelligent design is "not science" because "the ideas that form the basis" of it "have never been tested by any scientific peer scrutiny or peer review." In 2005, the American Astronomical Society declared: "Intelligent design fails to meet the basic definition of a scientific idea: its proponents do not present testable hypotheses and do not provide evidence for their views."[1] And the Biophysical Society adopted a policy stating: "What distinguishes scientific theories" from intelligent design "is the scientific method, which is driven by observations and deductions." Since intelligent design is "not based on the scientific method," it is "not in the realm of science."[2]

But the definition of "science" is more controversial than these statements imply, intelligent design is both testable and tested, and what really matters is not whether a hypothesis fits a particular definition of "science" but whether we have good reason for thinking it's true.

Guess what?

🦶 Darwinists have been unable to refute intelligent design with evidence, so they rely on a self-serving definition of science that excludes it from serious consideration.

🦶 Intelligent design is as scientific as Darwinism unless "science" is arbitrarily defined to permit only natural explanations.

🦶 Darwinists also dismiss intelligent design by citing the "consensus" of scientists—but in the Middle Ages the "consensus" was that the sun revolved around the Earth!

What Is Science?

In 1982, U.S. District Judge William A. Overton declared unconstitutional an Arkansas law requiring public schools to give "balanced treatment" to biblical creationism and Darwinism. The judge ruled that the former is religion rather than science, based on five "essential characteristics of science" that had been presented to him by various witnesses—most notably philosopher Michael Ruse. Science, the judge ruled, is (1) guided by natural law; (2) based on explaining by natural law; (3) testable; (4) tentative; and (5) falsifiable.[3]

As we have seen, intelligent design is not the same as biblical creationism, so Overton's criteria might not apply to ID in the same way he thought they applied to biblical creationism. But are his criteria even valid?

Philosopher of science Larry Laudan rejects biblical creationism on empirical grounds, but he also rejects Overton's ruling that it is unscientific. Regarding the first two criteria, Laudan points out that scientists often make claims before they can explain them by natural law. For example: "Galileo and Newton took themselves to have established the existence of gravitational phenomena, long before anyone was able to give a

Trivial pursuit

"It is hard to see how anything like a reasonably serious dispute about what is and isn't science could be settled just by appealing to a definition. One thinks this would work only if the original query were really a verbal question...[but] the question would be trivial, like the question whether there are any married bachelors."

—Philosopher **Alvin Plantinga**

causal or explanatory account of gravitation." And although Laudan considers the creationist claim of a worldwide flood to be false, he argues that to suggest such a claim "is unscientific until we have found the laws on which the alleged phenomenon depends is simply outrageous."[4]

Regarding the third and fifth criteria, some philosophers of science have suggested two ways to test a hypothesis. One is to seek evidence *for* it ("verification"), and the other is to seek evidence *against* it ("falsification"). Laudan points out that "the vast majority of non-scientific and pseudoscientific systems of belief have verifiable constituents." For example, astrologers sometimes make verifiably true predictions, yet most scientists regard astrology as unscientific. On the other hand, falsifiability "has the untoward consequence of countenancing as 'scientific' every crank claim which makes ascertainably false assertions."[5]

So although everyone agrees that scientific hypotheses should be tested against the evidence, verifiability and falsifiability are not reliable criteria to distinguish science from non-science. In recent years, Darwinists have increasingly relied on another criterion: "methodological naturalism." Science, they say, can invoke only natural explanations. As Ruse testified: "Science attempts to explain the empirical world in terms of natural law and natural processes."[6]

Methodological Naturalism

Advocates of this position distinguish between "methodological" and "metaphysical" naturalism. The latter is materialistic philosophy—the view that nature is all that exists, and God and spirit are imaginary. The former is supposedly just a statement about an inherent limitation on the scientific method rather than a description of the whole of reality.

Nevertheless, *defining* science as the search for natural explanations leads to serious problems. First, miracles might really happen. Laboratory experiments generally focus on phenomena that obey natural regularities, for which appeals to miracle would be inappropriate. But if God *does* work miracles, whether in the present or in the past, we would be seriously mistaken to insist that they can be explained by natural regularities. Acts of God that take place in the real world would have objective effects, and if methodological naturalism prevents us from even considering an essential element in their causation, but claims to lead to the truth, then it is the same as metaphysical naturalism.

According to philosopher Alvin Plantinga: "If you exclude the supernatural from science, then if the world or some phenomena within it are supernaturally caused—as most of the world's people believe—you won't be able to reach that truth scientifically. Observing methodological naturalism thus hamstrings science by precluding science from reaching what would be an enormously important truth about the world. It might be that, just as a result of this constraint, even the best science in the long run will wind up with false conclusions."[7]

A Book You're Not Supposed to Read

Nature, Design, and Science: The Status of Design in Natural Science, by Del Ratzsch; State University of New York Press, 2001.

Second, methodological naturalism may induce people to cling to ideas that are unsupported—or actually contradicted—by the evidence. If a person refuses to question the doctrine of universal common ancestry simply because it is the best naturalistic explanation available, even though fossils cannot provide evidence for it and the molecular evidence is inconsistent with it—at least at the level of the major domains of life—then that person is no longer engaged in the activity most of us think of as science: namely, determining which hypotheses best fit the evidence. The same could be said of a person who insists that macroevolution is just an extrapolation of microevolution simply because that is the best naturalistic explanation, even though the evidence from living organisms strongly suggests that it's not true. As Darwin critic Phillip E. Johnson has pointed out: "Naturalism and empiricism are often erroneously assumed to be very nearly the same thing, but they are not. In the case of Darwinism, these two foundational principles of science are in conflict."[8]

Of course scientists should seek natural causes. According to Dembski, that is the first step in making a design inference. His explanatory filter (Chapter Eight) leads to design only after undirected natural causes such as necessity and chance are eliminated. The problem with methodological naturalism is that it arbitrarily prohibits an investigator from moving beyond undirected natural causes. According to philosopher Del Ratzsch: "If one restricts science to the natural, and assumes that science can in principle get to all truth, then one has implicitly assumed philosophical naturalism . . . Methodological naturalism is not quite the lamb it is sometimes pictured as being."[9]

Criteria based on natural law, testability, and methodological naturalism are not only questionable, but they can also be used to declare Darwinism unscientific. First, as we saw in Chapter Six, Darwinists claim that universal common ancestry is a fact whether or not we know the natural laws underlying descent with modification. Second, as we saw in Chapters Two and Four, universal common ancestry is not testable by the fossil or molecular evidence, and as stated in Chapter Five, the Darwinian origin of species may be untestable too. Third, as we saw above, methodological naturalism leads Darwinists to tolerate empirical anomalies that would sink most scientific theories. Since natural law, testability, and methodological naturalism are inadequate, some Darwinists try to rule out intelligent design on sociological and psychological grounds.

Sociological and Psychological Criteria

As we saw in Chapter Six, Darwinists often appeal to the "consensus" of scientific opinion, even though a scientific consensus is notoriously unreliable. One thing the history of science shows us is that the scientific consensus at any given time is almost certain to change. In fact, the growing number of scientists today who are skeptical of Darwinism shows that the current consensus is changing before our eyes.

As illustrated in Chapter Nine, Darwinists also rely heavily on the fact that ID theorists seldom publish in peer-reviewed science publications. (Harvey Lodish's claim that ID has "never" been peer reviewed is false.)[10] Since most science publications are openly committed to promoting Darwinism, the current scarcity of peer-reviewed articles defending intelligent design is not surprising. Indeed, many scientific theories that are now widely accepted were initially rejected by most of their authors' peers—including Darwinian evolution.

So the sociological fact that a majority of the scientific community currently rejects intelligent design is not a reliable indicator of ID's scientific status or future prospects. To bolster their rejection of ID, Darwinists bring in psychological criteria as well. These include questions of personal religious motivation, the alleged religious implications of intelligent design, and Judge Overton's fourth criterion: tentativeness.

First, Darwinists claim that ID theorists are motivated by religious commitments. But as a blanket statement this is manifestly untrue; otherwise, noted British atheist Antony Flew would not have become convinced in 2004 that the evidence from nature points to design. On the other hand, many Darwinists are apparently motivated by a personal commitment to atheism. Richard Lewontin insists "we cannot allow a Divine Foot in the door"; Barbara Forrest opposes efforts "to explain the world in supernatural terms." If personal religious motivations are used to test the scientific status of a hypothesis, Darwinism itself fails the test.[11]

Second, Darwinists complain that intelligent design is not scientific because it has religious implications. It is true that most (though not all) ID theorists believe that the designer of the cosmos and living things is the God of the Bible, but they also acknowledge that this belief goes beyond the evidence of design in nature. On the other hand, many Darwinists believe that their hypothesis *excludes* God. Tufts University philosopher Daniel C. Dennett calls Darwinism a "universal acid" that "eats through just about every traditional concept"—especially the concept of God. Richard Dawkins has famously written that "Darwin made it possible to be an intellectually fulfilled atheist." Cornell University evolutionary biologist William B. Provine calls Darwinism "the greatest engine of atheism ever invented," since it shows that "no gods worth having exist." Obviously, for many people Darwinism has implications for religious belief, so by this criterion Darwinism is no more scientific than ID unless "science" is arbitrarily equated with materialistic philosophy.[12]

Finally, some Darwinists claim that ID is not scientific because its advocates hold to it dogmatically rather than tentatively. As we have seen, though, many people hold to Darwinism dogmatically in the face of anomalous evidence from the fossil record, embryology, and molecular analyses. In any case, personal dogmatism is irrelevant to determining the empirical justification for a theory. As Laudan puts it, when "several experiments turn out contrary to the predictions of a certain theory, we do not care whether the scientist who invented the theory is prepared to change his mind."[13]

In 1938, German physicist Carl F. von Weizsächer gave a talk in which he referred to the relatively new idea of the Big Bang. Renowned physical chemist Walther Nernst, who was in the audience, became very angry. Weizsächer later wrote: "He said the view that there might be an age of the universe was not science. At first I did not understand him. He explained that the infinite duration of time was a basic element of all scientific thought, and to deny this would mean to betray the very foundations of science. I was quite surprised by this idea and I ventured the objection that

A Book You're Not Supposed to Read

Debating Design: From Darwin to DNA, edited by William A. Dembski and Michael Ruse; Cambridge University Press, 2004.

it was scientific to form hypotheses according to the hints given by experience, and that the idea of an age of the universe was such a hypothesis. He retorted that we could not form a scientific hypothesis which contradicted the very foundations of science." Weizsäcker concluded that Nernst's reaction revealed "a deeply irrational" conviction that "the world had taken the place of God, and it was blasphemy to deny it God's attributes."[14]

So Nernst wrongly—and dogmatically—considered the Big Bang unscientific on the basis of his own personal beliefs. Like natural law, verifiability, falsifiability, and methodological naturalism, psychological criteria cannot adequately describe the scientific enterprise. As Laudan wrote after the Arkansas decision: "There is no demarcation line between science and non-science, or between science and pseudoscience, which would win assent from a majority of philosophers. Nor is there one which *should* win acceptance from philosophers or anyone else."[15]

The Science of Intelligent Design

Criteria for demarcating science from non-science fail to resolve the controversy between Darwinism and intelligent design, not only because the criteria themselves are flawed but also because intelligent design fulfills many of them at least as well as Darwinism does.

Consider testability. Despite problems with specific criteria such as verifiability and falsifiability, everyone agrees that scientific hypotheses should somehow be tested against evidence from the natural world. According to design theorists, we can determine from empirical evidence whether some features of the natural world are better explained by an intelligent cause than by unguided natural processes. This is not natural theology, since it does not purport to prove the existence and attributes of God. It does not claim that all features of the natural world are designed (though it does not rule out that possibility, either). It does not begin with, or claim to defend, the Bible or any religious doctrine. It is merely a statement that we can infer intelligent design from its effects.

According to ID theorists, those effects in living things include irreducible complexity (Behe) and complex specified biological information (Meyer). The hypothesis of irreducible complexity can be tested: if a single system composed of several well-matched interacting parts continues to function when any one of the parts is removed, then that system is not irreducibly complex. Evolutionary biologists Jerry A. Coyne, Russell F. Doolittle, and Kenneth R. Miller all claim to have provided evidence that disproves Behe's hypothesis in specific cases. If irreducible complexity were unscientific because it is untestable, how could Coyne, Doolittle, and Miller all claim to have tested it and shown it to be false? A hypothesis cannot be both untestable and tested.

The same can be said for Meyer's hypothesis that biological information cannot be produced by unguided natural processes. Miller claims that evidence supporting the "RNA world" scenario for the origin of life disproves Meyer's hypothesis. Meyer disagrees, but the mere fact that

From *Seinfeld*: Kramer has an altercation with a monkey at the zoo

Kramer: So, so what do you want me to do?

Mr. Pless (zoo official): Well, frankly, we'd like you to apologize.

Kramer: Yeah, well, he started it.

Mr. Pless: Mr. Kramer, he is an innocent primate.

Kramer: So am I. What about my feelings? Don't my feelings count for anything? Oh, only the poor monkey's important. Everything has to be done for the monkey!

Miller bases his claim on evidence shows that Meyer's hypothesis is testable.

In cases such as these, Darwinists often complain that intelligent design consists of nothing more than criticisms of Darwinism. To be sure, part of the process of inferring design involves eliminating explanations based on natural law and chance, as we saw above. Since Darwinists claim that their explanations disprove intelligent design, it is reasonable for ID theorists to rebut them. But ID theorists do much more than argue against the alleged power of natural selection and random variations; they also argue that the features in question have characteristics that we normally attribute to design. Since irreducible complexity is normally caused by an intelligent agent that assembles parts with a future function in mind, and since the only known cause of complex specified information is intelligence, a design inference is warranted in these cases.

Future research may or may not corroborate ID, but there can be no doubt that ID and Darwinism are looking at the same evidence and giving different answers to the same questions. Darwinists attempt to insulate their answers from criticism by declaring ID unscientific, but their attempt collapses into a contradiction: ID isn't science because it isn't testable, and, besides, it has been tested and proven false.

But Is It True?

Clearly, definitions of "science" that depend on natural law, testability, motivations, implications, or tentativeness cannot discriminate between Darwinism and intelligent design. The only definition of "science" that would include Darwinism and exclude intelligent design would be that "science" allows only natural explanations. But unless this exclusion is accompanied by an acknowledgment that some objective phenomena might not have natural explanations, then it is simply materialistic philosophy. It is not based on evidence; indeed, *it has no need for evidence—*

except perhaps as window dressing. The only other way Darwinism can trump ID is by relying on a "scientific consensus." But no reasonable person thinks that science is just the majority opinion of those who call themselves scientists at any given time. If it were, scientists would still believe that the sun goes around the Earth.

Plantinga asks whether Darwinists "really mean to suggest that the dispute can be settled just by looking up the term 'science' in the dictionary? If so, they should think again." It is "hard to see how anything like a reasonably serious dispute about what is and isn't science could be settled just by appealing to a definition. One thinks this would work only if the original query were really a verbal question—a question like... whether there are any married bachelors." Plantinga concludes: "The real question, I think, lies in a quite different direction."[16]

The real question is whether we have good evidence-based reasons to think a hypothesis is true. Laudan distinguishes between "What makes a belief well founded (or heuristically fertile)? And what makes a belief scientific?" He concludes that our focus "should be squarely on the empirical and conceptual credentials for claims about the world. The 'scientific' status of those claims is altogether irrelevant." According to

Highly irrelevant

"There are two distinct issues: What makes a belief well founded (or heuristically fertile)? And what makes a belief scientific?...Our focus should be squarely on the empirical and conceptual credentials for claims about the world. The 'scientific' status of those claims is altogether irrelevant."

—Philosopher of science **Larry Laudan**, 1983

Laudan, debating the scientific status of a hypothesis merely diverts attention away from whether the evidence provides stronger support for Darwinian evolution than for its competitors. "Once that question is settled, we will know what belongs in the classroom and what does not."[17]

Chapter Thirteen

❖❖❖❖❖❖❖❖❖❖❖❖❖

TO TEACH, OR NOT TO TEACH

D espite the alarmist rhetoric of some Darwinists, there are no organized efforts in the United States—not even by biblical creationists—to ban Darwinism from public school science classrooms. As we have seen, there are serious problems with the evidence for Darwinism, and a significant number of scientists acknowledge this fact. Should biology teachers be permitted, encouraged, or even required to inform students of those problems?

A separate question is: should teachers be permitted, encouraged, or required to include intelligent design in biology classes as a possible alternative to Darwinism? Although some Darwinists claim that teaching the scientific controversy over Darwinism is the same as teaching intelligent design, it clearly is not.

A Tale of Two Teachers

About an hour and a half north of Seattle lies Burlington, Washington. In May 1998, the local newspaper carried a front-page story titled "Teacher's approach reignites debate on evolution." For several years, biology teacher Roger DeHart had been teaching students the required curriculum about evolution, but he had also been mentioning intelligent design. Asked about the practice, DeHart said, "I'm just saying, 'Here's another

Guess what?

🦶 Teaching students the evidence for and against Darwinism is not the same as teaching intelligent design.

🦶 The U.S. Congress has officially endorsed teaching students "the full range of scientific views" about Darwinian evolution.

🦶 When Ohio adopted science standards recommending a critical analysis of evolution, the Darwinists waged a relentless campaign until the critical analysis was dropped. They are now trying to do the same in Kansas.

143

way of interpreting the issue, so that you understand the issues well enough to know both sides.' My personal opinion should stay out of this." Some of DeHart's former students confirmed that "he was fair to both sides and offered no personal opinions."[1]

DeHart had the support of his school administrators and local school board, but the American Civil Liberties Union (ACLU) claimed that his practice amounted to religious proselytizing that "violates both state and federal laws." The school district caved in to pressure from the ACLU, the National Center for Science Education (NCSE), and local atheists.

Congress weighs in

"The Conferees recognize that a quality science education should prepare students to distinguish the data and testable theories of science from religious or philosophical claims that are made in the name of science. Where topics are taught that may generate controversy (such as biological evolution), the curriculum should help students to understand the full range of scientific views that exist, why such topics may generate controversy, and how scientific discoveries can profoundly affect society."

Conference Report, No Child Left Behind Act of 2001

DeHart was ordered to stop mentioning intelligent design, though he was told in 2001 that he could request approval to use supplementary materials critical of some of the evidence for Darwinian evolution.

When DeHart submitted such materials, however—from mainstream science publications such as *American Biology Teacher*, *Natural History*, *The Scientist*, and *Nature*—he was denied permission to use or even mention them. He was reassigned to a different subject, and his biology class was assigned to a physical education instructor right out of teacher training. In 2002, DeHart left his career as a public high school teacher and moved with his wife and children to another state.[2]

In the meantime, another public high school biology teacher in University Place, Washington—about an hour southwest of Seattle—was having a very different experience. To prepare his students to be "competent jurors" on Darwinism, Doug Cowan taught them "more than they need[ed] to know about evolution." Specifically, Cowan taught everything in the required curriculum on Darwinian evolution—including the reputable evidence for it—but he also informed his students that "some current biology textbooks contain widely discredited evidence" for Darwinism, at which point, he said, "the last of the sleepy looks in the classroom usually vanishes."

Cowan did not include intelligent design in his lessons, though he retained "the right to mention the viewpoint of design theory if asked by curious students if there are other explanations out there in the world of scientific thinking." His principal and district superintendent backed him, on the condition that he remain neutral in his presentation. As a testimony to his neutrality, Cowan says that "after my presentations, many kids will ask me what I believe, since they cannot tell what my position is."[3]

Cowan, unlike DeHart, still has his job. One reason, no doubt, is that in 2001 the U.S. Congress endorsed teaching of the controversy over Darwinism.

Teaching the Controversy

As Congress was debating the No Child Left Behind Act of 2001, Republican senator Rick Santorum of Pennsylvania proposed a "Sense of the Senate" amendment that subsequently became part of the joint House-Senate Conference Report accompanying the final bill. "The Conferees recognize," stated the report, "that a quality science education should prepare students to distinguish the data and testable theories of science from religious or philosophical claims that are made in the name of science. Where topics are taught that may generate controversy (such as biological evolution), the curriculum should help students to understand the full range of scientific views that exist, why such topics may generate controversy, and how scientific discoveries can profoundly affect society."[4]

Darwinists claim that the Santorum amendment is irrelevant because it was not part of the actual legislation. To counter this misinformation, Republican congressman John A. Boehner of Ohio, Republican senator Judd Gregg of New Hampshire, and Santorum wrote in 2003:

> This statement was included in HR 107-334, which was approved by both houses of Congress in December 2001. It therefore represents the official view not only of the Conference Committee but of the United States Congress as a whole about how science instruction should proceed under the No Child Left Behind Act. . . . Report language provides official guidance from Congress on how statutory language should be enforced by other government agencies, and the Santorum language should be understood in this light . . . [It] made explicit Congress's rejection of the idea that students only need to learn about the dominant scientific view of controversial topics. The Santorum language clarifies that public school students are entitled to learn that there are differing scientific views on issues such as biological evolution.[5]

Many Darwinists refuse to acknowledge that there are "differing scientific views" on biological evolution. In 2000, Eugenie C. Scott claimed that "evidence against evolution is just a euphemism for creation science." In 2003, Scott and Glenn Branch (both officials of the NCSE—the same organization that helped to destroy DeHart's public school teaching career) wrote in the journal *Trends in Ecology and Evolution* that "students are unlikely to be able to understand both sides of the controversy," and "although 'teaching the controversy' sounds fair, it is unfair to pretend to students that a controversy exists in science where none does."[6]

In the same journal in 2004, Stephen C. Meyer pointed out that Scott herself has acknowledged that there is "significant scientific debate" about the sufficiency of the neo-Darwinian mechanism of natural selection acting on random mutations, and that "many evolutionary biologists now disagree" over universal common descent. Meyer concluded: "Teaching students about scientific controversies is less a matter of fairness . . . than it is a matter of full scientific disclosure." Scott and Branch replied that the controversies are much narrower than Meyer implied, and they concluded by questioning his "motivations, agenda, and methods."[7]

But Meyer's alleged motivations are irrelevant. As we have seen in previous chapters, Darwinism has serious problems with the evidence, and some highly qualified scientists are skeptical of its exaggerated claims. Shouldn't science students be permitted to learn about them?

Of course, this is not the same as teaching intelligent design. If Darwinists are opposed to mentioning scientific problems with their view, you would think they would be even more opposed to mentioning intelligent design. Yet Darwinists have been discussing ID in public school science classes for years.

Teach ID, but Do It Badly

Biology textbooks have been mentioning intelligent design since the late 1990s—but only to misrepresent and disparage it. In his 1998 *Evolutionary Biology* Douglas J. Futuyma wrote that biblical creationists "adopted the camouflage" of "teaching that the complexity of living things can be explained only by intelligent design." The 2002 edition of Peter H. Raven and George B. Johnson's *Biology* claimed that "the intelligent design argument" is "easily answered," since "complex structures like eyes evolved as a progression of slight improvements." The 2004 edition of Sylvia S. Mader's *Biology* asked: "Should schools

A Book You're Not Supposed to Read

Darwinism, Design, and Public Education, edited by John Angus Campbell and Stephen C. Meyer; East Lansing, MI: Michigan State University Press, 2003.

be required to teach an intelligent-design theory that traces its roots back to the Old Testament and is not supported by observation and experimentation?" So Raven and Johnson merely restated Darwin's theory of evolution through slight improvements, while Futuyma and Mader misrepresented intelligent design as a disguised form of biblical creationism.[8]

In his 2005 college textbook *Evolution,* Futuyma writes that intelligent design "is not testable . . . ; it cannot be evaluated by the methods of science." A few pages later, however, he argues that evidence of intermediate stages in what are supposedly irreducibly complex structures provide a "refutation of the intelligent design position." He can't have it both ways: intelligent design is either untestable, or it's been tested and proven false.[9]

Central Washington University biology professor Steven H. Verhey reported in the journal *BioScience* in 2005 that he assigned his students an "ID text in parallel with a critique of the book and a well-known defense of evolution." Verhey measured his success at helping students "to arrive at mature views of the role of evolution in biology" by asking how far he was able to move them along a scale of beliefs from the "nonrationalist Christian literalist" end to the "rationalist atheistic evolutionist" end.[10]

But why is a biology professor at a public university trying to convert students to atheism? And why did he use as his "ID text" my book *Icons of Evolution: Why Much of What We Teach About Evolution Is Wrong*, which criticizes distorted evidence for Darwinism (such as Haeckel's embryo drawings) but is not about intelligent design at all? (The book mentions ID only in two sentences describing Roger DeHart's ordeal.) Of course, it's easier to convince students that ID is wrong when they aren't assigned any materials that actually defend it.

A Book You're Not Supposed to Read

Law, Darwinism, and Public Education, by Francis J. Beckwith; Lanham, MD: Rowman and Littlefield Publishers, 2003.

Clearly, Darwinists approve of mentioning intelligent design in science classes—as long as it is misrepresented and ID theorists are given no opportunity to respond.

Kansas and Ohio

In 1999, the Kansas State Board of Education revised its science standards. A pro-Darwin writing committee proposed a substantial increase in the space devoted to evolution and insisted that students "understand" that large-scale evolutionary changes are explained by natural selection and genetic changes. Several members of the board wanted to include some acknowledgement of the scientific controversy over macroevolution, but the pro-Darwin board members refused. The resulting compromise increased the space devoted to evolution, but it included only microevolution. Unfortunately, the board also deleted references to the Big Bang.[11]

Kansas Darwinists were furious, and they misinformed the news media that the board had eliminated evolution entirely. Some reports even claimed—falsely—that Kansas had prohibited the teaching of evolution or mandated the teaching of biblical creationism. Herbert Lin of the National Research Council wrote that American colleges and universities should declare "their refusal to count as an academic subject any high school biology course taught in Kansas," and *Scientific American* editor John Rennie wrote: "Send a clear message to the parents in Kansas that this bad decision carries consequences for their children." Unless Kansans surrendered, Darwinists threatened to hold their kids hostage.[12]

In the next school board election, pro-Darwin candidates won a majority of seats on the Kansas board and revised the state standards in 2001 to include macroevolution—with no mention of the scientific controversy over it. Then Ohio entered the fray.

In 2002, the Ohio State Board of Education debated whether to revise its science standards to include intelligent design as an alternative to

Darwinian evolution. Darwinist Lynn Elfner of the Ohio Academy of Science predicted "there would be a major revolt" by Ohio scientists if ID were included. Board of Education member Martha Wise said "it's a shrouded way of bringing religion into the schools." *Slate* correspondent William Saletan wrote that "the Taliban has invaded Ohio" in the form of intelligent design advocates.[13]

To hear from both sides of the controversy, the board decided to hold a public debate on March 11, 2002, between me and Stephen C. Meyer from the Discovery Institute on one side, and Case Western Reserve University physicist Lawrence M. Krauss and Brown University biologist Kenneth R. Miller on the other. In the debate, however, Meyer and I did *not* argue for mandating intelligent design in the state standards—though we both consider ID fully scientific. We proposed that students be taught the scientific evidence and arguments for and against Darwinism, and that Ohio should not enact standards that would prevent the discussion of other theories.[14]

In December 2002, the Ohio State Board of Education adopted new science standards that included a benchmark stating: "Explain the historical and current scientific developments, mechanisms, and processes of biological evolution. Describe how scientists continue to investigate and critically analyze aspects of evolutionary theory. (The intent of this benchmark does not mandate the teaching or testing of intelligent design.)"[15]

Darwinists in the news media portrayed this as biblical creationism. Toledo *Blade* associate editor Eileen Foley called those who voted for a critical analysis of evolution "know-nothing fanatics...intent on twisting science to accommodate their pet biblical phrases." When the board voted in March 2004 to adopt a model curriculum to implement the critical analysis, the *Columbus Dispatch* reported: "State OKs curriculum involving creationism," and *Cleveland Plain Dealer* columnist Sam Fulwood called its supporters "knuckle-dragging creationists" who were waging a "state-by-state jihad against real science."[16]

Despite the Darwinists' campaign of disinformation and character assassination, the Ohio board stood its ground. In 2005, the center of the storm moved back to Kansas again. The Darwinists had lost their majority on the Kansas State Board of Education, which decided to take another look at the science standards. A writing committee proposed increasing the emphasis on Darwinism, but a significant minority of the committee wanted to include some critical analysis of it as well.

The Kansas board decided to hold public hearings, with each side given three days to present expert witnesses who would be cross-examined by attorneys from the other side. The Darwinists—who in 2002 had effectively lost the debate in Ohio—refused to cooperate. Liz Craig, of the militantly pro-Darwin Kansas Citizens for Science, wrote to her colleagues: "My strategy at this point is the same as it was in 1999: notify the national and local media about what's going on and portray them in the harshest light possible, as political opportunists, evangelical activists, ignoramuses, breakers of rules, unprincipled bullies, etc."[17]

By April 2005, the Darwinists had organized a boycott of the proposed hearings. As *Nature* put it, "Biologists snub 'kangaroo court' for Darwin." Alan I. Leshner, chief executive officer of the pro-Darwin American Association for the Advancement of Science (AAAS), wrote to the Kansas State Department of Education: "AAAS respectfully declines to participate in this hearing, out of concern that rather than contribute to education, it will most likely serve to confuse the public about the nature of the scientific enterprise. The fundamental structure of the hearing suggests that the theory of evolution may be debated."[18]

FIGURE 8. Ohio high school science teacher Bryan Leonard, who helped draft the model curriculum implementing Ohio's critical analysis of evolution in 2004, testifying before the Kansas State Board of Education in 2005.

Photo courtesy of www.NewLibertyVideos.com.

The hearing took place anyway, in May 2005. Twenty-three critics of Darwinism testified, including seventeen with doctoral degrees. Ten witnesses were university science professors, two were philosophy professors, and three were high school teachers. One of the high school teachers was Bryan Leonard, who had helped to draft the model curriculum for Ohio's critical analysis of evolution in 2004 (Figure 8).[19]

A central issue in the hearings was the definition of science. In 2001, the pro-Darwin board had changed the definition of science to read: "Science is a human activity of systematically seeking natural explanations for what we observe in the world around us." In 2005, the newly constituted board proposed to change this to read: "Science is a systematic method of continuing investigation that uses observation, hypothesis test-

Hail Darwin!

"Teachers seeking to 'teach the controversy' over Darwinian evolution in today's climate will likely be met with false warnings that it is unconstitutional to say anything negative about Darwinian evolution. Students who attempt to raise questions about Darwinism, or who try to elicit from the teacher an honest answer about the status of intelligent design theory will trigger administrators' concerns about whether they stand in Constitutional jeopardy. A chilling effect on open inquiry is being felt in several states already, including Ohio, South Carolina, and Pennsylvania. [District Court] Judge Jones's message is clear: give Darwin only praise, or else face the wrath of the judiciary."

Traipsing into Evolution: Intelligent Design and the Kitzmiller vs. Dover Decision, by **David K. DeWolf**, **John G. West**, **Carey Luskin**, and **Jonathan Witt**; Seattle, WA: Discovery Institute Press, 2006.

ing, measurement, experimentation, logical argument, and theory building to lead to more adequate explanations of natural phenomena." The Darwinists objected that science must be limited to natural explanations—though as we saw in Chapter Twelve, in practice this usually amounts to materialistic philosophy. The Darwinists also claimed that the proposed change would be a "radical redefinition" of science—though a survey of science standards in other states revealed that only Kansas had defined science as the search for natural explanations. It was the Darwinists, not their critics, who had radically redefined science in 2001. The 2005 board was merely trying to bring Kansas back into line with established educational practice in other states.[20]

Although Darwinists refused to testify at the hearing, Topeka attorney Pedro Irigonegaray announced that he would cross-examine the witnesses who testified in favor of a critical analysis of evolution. At the end of the proceedings, Irigonegaray gave a two-hour speech defending Darwin-only education and denouncing the Kansas State Board of Education. Breaking the procedural rules announced at the beginning of the hearing, Irigonegaray refused to be cross-examined.[21]

On November 8, 2005, the board adopted science standards that included provisions for the critical analysis of evolution. True to form, the AAAS stated that "we are deeply disturbed by the vote . . . to mix science and faith in public school science classrooms," claiming that it put "the children of Kansas" at "great risk."[22]

Meanwhile, back in Ohio, the Darwinists had not given up in their relentless efforts to eliminate the critical analysis of evolution from that state's science curriculum. In February 2006, the Ohio State Board of Education finally caved in to the Darwinists and deleted it. The move was spearheaded by board member Martha Wise, who reportedly once claimed that God told her that critical analysis of evolution was wrong. On February 22, 2006, she wrote: "I believe in God the creator. I believe in freedom, I believe in America, and the state of Ohio, and the Republican

Party, fiscal conservatism, fairness, and honesty. These values guided me last week to lead the Ohio Board of Education to remove creationism from our state's Science Standards."[23]

So Ohio's critical analysis, which specifically stated that it did not require the teaching of intelligent design, much less creationism, was removed largely through the efforts of a woman who claims she was guided by God to remove creationism from the science standards! For some strange reason, the separation-of-church-and-state lobby had no problem with this. Imagine the reaction if a board member had said that God guided her to *support* the critical analysis of evolution—or worse, intelligent design. A lawsuit would have followed as surely as night follows day.

What had changed between 2002 and 2006 to make the Ohio State Board of Education cave in to Darwinist bullies? The single most important factor was a December 2005 decision by a federal judge in Pennsylvania.

Traipsing into Evolution

In November 2004, the Dover, Pennsylvania, school board adopted a policy requiring that a statement about evolution and intelligent design be read to public school biology students about to spend a week studying Darwinism. The statement read, in part: "Because Darwin's Theory is a theory, it continues to be tested as new evidence is discovered. The Theory is not a fact. Gaps in the Theory exist for which there is no evidence. A theory is defined as a well-tested explanation that unifies a broad range of observations. Intelligent Design is an explanation for the origin of life that differs from Darwin's view. The reference book *Of Pandas and People* is available for students who might be interested in gaining an understanding of what Intelligent Design actually involves. With respect to any theory, students are encouraged to keep an open mind."[24]

Some major proponents of ID—in particular, the Discovery Institute—urged the board to rescind the policy. The Discovery Institute defends the right of teachers to discuss ID if they choose to do so, but it opposes *requiring* ID until it becomes better established in the scientific community.[25] But the Dover School Board ignored the Discovery Institute's advice. The American Civil Liberties Union (ACLU)—the same organization that defends the right of Nazis to march publicly in support of their racist and anti-Semitic policies—took the school district to court, claiming that a five-minute statement encouraging critical thinking is too much for students about to be subjected to a week of indoctrination in Darwinism.

In the ensuing trial, Darwinists testified that intelligent design is not science because it is not testable. Besides, it has been tested and found to be false. Members of the Dover school board admitted that they were influenced by their religious beliefs, but the judge ignored testimony that several of the Darwinists were influenced by other religious beliefs, including atheism. Judge John E. Jones III was so impressed by the testimony and materials presented by the Darwinists that he apparently didn't bother to read much of the material presented by their critics, and he came down squarely on the side of the ACLU.

Apparently not burdened with an excess of judicial restraint, Judge Jones wrote in his decision: "The Court is confident that no other tribunal in the United States is in a better position than we are to traipse into this controversial area." He then concluded: "ID is not science and cannot be adjudged a valid, accepted scientific theory as it has failed to publish in peer-reviewed journals, engage in research and testing, and gain acceptance in the scientific community."

On the tube

From *Friends*: Ross on evolution

Ross: Okay, Pheebs. See how I'm making these little toys move? Opposable thumbs. Without evolution, how do *you* explain opposable thumbs?

Phoebe: Maybe the overlords needed them to steer their spacecrafts.

Ross: Please tell me you're joking.

Phoebe: What is this obsessive need you have to make everyone agree with you? No, what's that all about? You know what I think? I think maybe it's time you put *Ross* under the microscope.

Ross: Is there blood coming out of my ears?

Judge Jones then issued a permanent injunction against the Dover School District, prohibiting it from requiring teachers to "denigrate or disparage the scientific theory of evolution," and from requiring them to mention ID.[26]

The infamous Scopes Monkey Trial was thus reversed with a vengeance. In 1925, the ACLU went to court to lift a prohibition against teaching Darwinism. Eighty years later, the ACLU persuaded a federal judge to prohibit teaching anything *but* Darwinism. Even though this prohibition has no legal force outside of central Pennsylvania, Ohio Darwinists prepared a handout titled "Direct Links between the Dover decision and Ohio's Creationist Lesson." With this, and Martha Wise's guidance from God, the Ohio State Board of Education was intimidated into surrendering its critical analysis of evolution. Ohio teachers, like teachers in central Pennsylvania, will henceforth be required to teach Darwinism, the whole of Darwinism, and nothing but Darwinism. Whatever happened to the truth, the whole truth, and nothing but the truth?

Chapter Fourteen

❧❧❧❧❧❧❧❧❧❧❧❧

DARWINISM AND CONSERVATIVES

O n May 25, 2006, Democrat-turned-Republican mayor of New York City Michael R. Bloomberg addressed graduates of the Johns Hopkins University School of Medicine in Baltimore. "Today," he said, "we are seeing hundreds of years of scientific discovery being challenged by people who simply disregard facts that don't happen to agree with their agendas. Some call it 'pseudoscience,' others call it 'faith-based science,' but when you notice where this negligence tends to take place, you might as well call it 'political science'." Bloomberg continued: "It boggles the mind that nearly two centuries after Darwin, and eighty years after John Scopes was put on trial, this country is still debating the validity of evolution." Then he falsely claimed that Kansas is "now proposing to teach 'intelligent design'—which is really creationism by another name—in science classes."[1]

Although Bloomberg did not specifically mention conservatives, there was no doubt that he was targeting them. The next day, the *New York Daily News* reported: "Mayor Bloomberg lashed out at conservatives yesterday for ignoring science and common sense on issues like ... evolution.[2]

Yet conservatives are far from unanimous in their opinions about Darwinism and intelligent design. Although most Darwinists are on the left end of the political spectrum and most intelligent design supporters are

Guess what?

❧ Although Darwinism finds most of its support on the left end of the political spectrum, some conservatives defend it as a biological basis for their views.

❧ A few people argue that Darwinism can serve as a foundation for morality, but historically it has been used to justify social evils such as eugenics and racism.

❧ Darwinian survival of the fittest presupposes a struggle for limited resources, inviting government regulation that hampers free market economies.

on the right, some conservatives are vocal opponents of ID. Some people even argue that Darwinism supports social and economic conservatism.

Conservatives against ID

On August 1, 2005, a newspaper reporter asked President George W. Bush for his "personal views" on "the growing debate over evolution versus intelligent design," and whether he thought "both should be taught in public schools." Bush replied that the decision should be left to local districts, but in his opinion "both sides ought to be properly taught . . . so people can understand what the debate is about."[3]

Darwinists were upset with the president's reply, but so was conservative columnist John Derbyshire. "This is Bush at his muddle-headed worst," he wrote, "conferring all the authority of the presidency on the teaching of pseudoscience in science classes." Derbyshire concluded: "What, then, should we teach our kids in high-school science classes? The answer seems to me very obvious. We should teach them *consensus* science," which means "we should teach them Darwinism, . . . the essential foundation for all of modern biology." We should also teach it "*unskeptically*, as settled fact," since teaching it otherwise "will only confuse young minds."[4]

Reviewing events in Pennsylvania and Kansas a few months later, conservative columnist Charles Krauthammer deplored "a fight over evolution that is so anachronistic and retrograde as to be a national embarrassment." Krauthammer called intelligent design "a fraud" and a "tarted-up version of creationism . . . whose only holding is that when there are gaps in some area of scientific knowledge—in this case, evolution—they are to be filled by God." Then, like Bloomberg, he falsely accused the Kansas Board of Education of "forcing intelligent design into the statewide biology curriculum."[5]

Ordinarily, conservatives such as Derbyshire and Krauthammer know better than to swallow propaganda peddled by the leftist-dominated news media, but in this case they seem to have let their guard down. As we saw in Chapter Seven, far from being "the essential foundation for all of modern biology," Darwinism largely claims credit for other scientists' work. And Derbyshire's insistence that Darwinism be taught "unskeptically, as a settled fact," without analyzing the actual evidence for and against it, is equivalent to insisting that it be taught as dogmatic philosophy rather than empirical science. Krauthammer has even less excuse for his misrepresentations of intelligent design and the Kansas situation. Apparently, he never bothered to read (or refused to believe) anything written by the people he criticizes.

Conservative columnist George Will also came out against intelligent design in November 2005. He criticized efforts by the Dover, Pennsylvania, school board "to insinuate religion, in the guise of 'intelligent design' theory, into high school biology classes"—because, Will claims, evolution is a fact. "Kansas's Board of Education," he wrote, "is controlled by the kind of conservatives who make conservatism repulsive to temperate people." Will concluded: "Limited-government conservatives will dissociate from a Republican Party more congenial to overreaching social conservatives."[6]

So Will regards the controversy over Darwinism and intelligent design as an embarrassing distraction from what he considers a more important issue: controlling federal spending. This is not unreasonable, but Will's ill-informed attacks on the way several states are struggling with Darwinist indoctrination in public school science classes don't help his case.

> ## The wit and wisdom of Hollywood
>
> "I believe in evolution and I think when it comes to business and the roots of business and the fundamentals of business, I don't think that ever changes. I think the idea of change is an illusion, but in nature it's necessary to change and perhaps business is a part of nature. I'm not totally sure."
>
> **—Elliott Gould**

Darwinian Conservatism?

Since Derbyshire, Krauthammer, and Will are unduly impressed by the propaganda favoring Darwinism, they want conservatives to stop resisting it. Northern Illinois University political science professor Larry Arnhart goes a step further by arguing that conservatives actually need Darwinism.

In 2005, Arnhart wrote: "Many conservatives fear Darwinism as promoting an atheistic and materialistic view of human life that is morally corrupting. Such fears should be dispelled by seeing how Darwinism actually provides a scientific understanding of the natural roots of morality and religion. . . . Instead of fearing science as the enemy of liberty, Darwinian conservatives can learn from science how liberty is founded in the nature of the human animal. That's why conservatives need Charles Darwin."[7]

Of course, many conservatives would respond that they don't fear science—far from it—but that they are skeptical of Darwinism's scientific pretensions. Since science is a good thing, and Arnhart equates science with Darwinism, it is not surprising that he wants to show how "modern

Immoral? No. Just no morals.

"Darwinian political theorists themselves cling tenaciously to moral aspirations—for nobility and universal justice—that cannot be adequately defended on the basis of their evolutionary naturalism." Indeed, Larry Arnhart's "Darwinian political theory... provides the basis for no useful moral teaching at all."

—Political scientist **Carson Holloway**, 2006

Darwinian science helps us to explain" how human nature "was formed as a product of natural evolution."[8] In order to do this, Arnhart relies on one of Darwinism's newer offshoots: sociobiology.

The most noted name in sociobiology is Harvard's Edward O. Wilson (who claims the word "Darwinism" was invented by creationists to discredit, well, Darwinism). Wilson defines altruism biologically as increasing another organism's ability to pass on genes at the expense of one's own, and he calls it the "culminating mystery of all biology."[9] It is a mystery because Darwinian evolution favors individuals who out-compete others to leave more offspring, so it is difficult or impossible for Darwinism to explain the existence of individuals who deliberately sacrifice for the sake of strangers. Yet if Darwinism cannot adequately explain altruism, how can it provide a foundation for morality and ethics?

Sociobiology suffers from other serious problems as well. Except for some rare pathological conditions, it has been impossible to tie human behavior to specific genes. (The "gay gene" that was much hyped a few years ago turned out to be a mirage.) If human behavior cannot be reduced to genetics, then according to neo-Darwinism it cannot be biologically inherited; if it cannot be biologically inherited, then it cannot evolve in a Darwinian sense. Still another problem with sociobiology is that it has been invoked to explain just about every human behavior from selfishness to self-sacrifice, from promiscuity to celibacy. A theory that explains something and its opposite equally well explains nothing.

It's no wonder that sociobiology and its latest manifestation, "evolutionary psychology" (called "evo-psycho" by some wags), are held in low regard even by some evolutionary biologists. Stephen Jay Gould once called sociobiology a collection of "just-so stories" in which "virtuosity in invention replaces testability as the criterion for acceptance." And in 2000 evolutionary biologist Jerry A. Coyne compared it to discredited Freudian psychology: "By judicious manipulation, every possible observation of human behavior could be (and was) fitted into the Freudian framework.

The same trick is now being perpetrated by the evolutionary psychologists. They, too, deal in their own dogmas, and not in propositions of science."[10]

Attempts to explain human behavior and values in Darwinian terms have been criticized not only by evolutionary biologists, but also by historians and political scientists. According to Oregon State University historian of science Paul L. Farber, biology "has been singularly unsuccessful in solving social problems or providing moral guidance," and efforts to base human values on evolutionary theory have "a dismal track record." University of Nebraska political science professor Carson Holloway contends: "Darwinian political theorists themselves cling tenaciously to moral aspirations—for nobility and universal justice—that cannot be adequately defended on the basis of their evolutionary naturalism." Holloway concludes that Arnhart's "Darwinian political theory... provides the basis for no useful moral teaching at all."[11]

Darwinism and Social Values

Far from providing support for traditional social values, Darwinism has repeatedly been used to subvert them. From a Darwinian perspective human beings are nothing more than highly evolved animals, so the same principles that breeders use with livestock should be applied to us. Thus Darwin wrote in *The Descent of Man*: "With savages, the weak in body or mind are soon eliminated; and those that survive commonly exhibit a vigorous state of health. We civilized men, on the other hand, do our utmost to check the process of elimination; we build asylums for the imbecile, the maimed, and the sick; ... Thus the weak members of civilized societies propagate their kind. No one who has attended to the breeding of domestic animals will doubt that this must be highly injurious to the race of man."[12]

This justified eugenics—the use of forced sterilization and infanticide to eliminate people deemed unfit by the cultural elite. Many early twentieth-century eugenicists in the U.S., Britain, and Germany explicitly based their deplorable policies on Darwinism.

One of the earliest and most prominent eugenicists was Ernst Haeckel, who used fake embryo drawings to illustrate what Darwin thought was "by far the strongest single class of facts" in favor of his theory. Haeckel also justified abortion on the grounds that human embryos before birth are merely lower forms of animals. A 1965 letter to the *New York Times* cited Haeckel's work to justify abortion. So did a 1981 letter from a prominent abortionist to the U.S. Senate.[13]

Darwinism also underwrote racism. Although racism predated Darwinism and many Darwinists oppose it, Darwin clearly regarded white Europeans as more highly evolved than other races. He predicted in *The Descent of Man* that at some future time "the civilized races of man will almost certainly exterminate, and replace, the savage races throughout the world." Since the higher apes will probably also be exterminated, "the break between man and his nearest allies will then be wider" because it will be between Caucasians and baboons "instead of as now between the Negro or Australian and the gorilla." Haeckel obligingly illustrated this point with a drawing in one of his books of an African man sitting in a tree with a chimpanzee, gorilla, and orangutan.[14]

In the first decade of the twentieth century, anthropologist Madison Grant was both head of the New York Zoological Society and a prominent eugenicist and racist. In 1906, he put a young Congolese pygmy named Ota Benga in the monkey house at the Bronx Zoo, alongside an orangutan, and exhibited him as an evolutionary missing link between apes and humans. He was freed only after some African American Baptist clergymen protested against the exhibit for promoting racism and Darwinism. Tragically, Benga later committed suicide.[15]

A Book You're Not Supposed to Read

From Darwin to Hitler, by Richard Weikart; New York: Palgrave Macmillan, 2004.

So Darwinism's impact on traditional social values has not been as benign as its advocates would like us to believe. Despite the efforts of its modern defenders to distance themselves from its baleful social consequences, Darwinism's connection with eugenics, abortion, and racism is a matter of historical record. And the record is not pretty.

It's the Economy, Stupid!

Some people argue that conservatives should embrace Darwinism because it justifies laissez-faire economics. According to this argument, Darwinism is good because it provides a biological basis for the competition necessary to free-market capitalism, and intelligent design is bad because it fosters a top-down regulation harmful to a free economy.

Darwin based his theory, in part, on Thomas Malthus's idea that populations increase faster than the resources that they need for survival. "A struggle for existence inevitably follows," wrote Darwin, resulting in what he called natural selection and what Herbert Spencer called survival of the fittest. Like Spencer, Darwin and his "bulldog" Thomas Henry Huxley used these ideas to defend laissez-faire economics. So did American sociologist William Graham Sumner.[16]

Darwinism and Hitler

"Darwinism by itself did not produce the Holocaust, but without Darwinism... neither Hitler nor his Nazi followers would have had the necessary scientific underpinnings to convince themselves and their collaborators that one of the world's greatest atrocities was really morally praiseworthy."

—Historian **Richard Weikart**

In 1944, historian Richard Hofstadter claimed that similar thinking motivated America's great business leaders. According to Hofstadter, John D. Rockefeller once said that "the growth of a large business is merely a survival of the fittest." In 1959, however, historian Irvin Wyllie showed that many of Hofstadter's claims were wrong. It was not successful businessmen but "scientists, social scientists, philosophers, clergymen, editors, and other educationally advantaged persons" who tended to be Darwinists. Most businessmen relied on the Bible, not *The Origin of Species*. The words Hofstadter put into the mouth of Bible-reading John D. Rockefeller had actually come from the industrialist's university-trained son.[17]

As political scientist John G. West has pointed out, nineteenth-century American businessmen and economists *did* praise free enterprise, competition, and laissez-faire capitalism. But they got their ideas from classical economists such as Adam Smith, not from Darwin. The same is true of most modern economists who have championed free enterprise. Ludwig von Mises argued that economic success depends on competition, but also on "mutual aid" and "social collaboration"—not a Darwinian struggle for existence. Friedrich A. von Hayek maintained that "complex and orderly" economic structures "might grow up which owed little or nothing to design" but "arose from the separate actions of many men who did not know what they were doing." Although he called this "the emergence of order as the result of adaptive evolution," Hayek emphasized that he was not talking about Darwinism. He wrote that concepts such as "natural selection," "struggle for existence," and "survival of the fittest" are "not really appropriate" in the social sciences.[18]

The Heart of the Matter

According to technology analyst George Gilder, the reason free enterprise works and excessive government control fails is that true wealth comes from human creativity and invention. "Because creativity is unpredictable," he wrote in 1981, "it is also uncontrollable. If the politicians want to have central planning and command, they cannot have dynamism and life. A managed economy is almost by definition a barren one, which can progress only by borrowing or stealing from abroad."[19]

Darwinism leads to managed economies, Gilder maintains, because it is a "zero-sum game." From a Darwinian perspective, limited resources inevitably provoke a struggle for existence that invites government regulation. But resources are *not* limited in the sense Darwinism assumes they are. Instead, they increase with new technological advances, technological advances come from human creativity, and creativity thrives in a free economy.

According to Gilder, Darwinism obstructs economic progress not only by encouraging government controls that inhibit creativity, but also by obscuring the very source of creativity. "The logic of creativity is 'leap before you look'," wrote Gilder in 1981. "Creative thought requires an act of faith. The believer must trust his intuition, the spontaneous creations of his mind, enough to pursue them laboriously to the point of experiment and knowledge.... It is love and faith that infuse ideas with life and fire. All creative thought is thus in a sense religious, initially a product of faith and belief.... God is the foundation of all living knowledge; and the human mind, to the extent it can know anything beyond its own meager reach, partakes of the mind of God."[20]

In response to a *Wired* magazine article defending Darwinism in 2004, Gilder wrote: "I believe that the notion that the intricate biological structures of the world bubbled up from a prebiotic brew and that ideas are an after-effect of a meaningless random material flux is the most sterile and stultifying notion in the history of human thought. It inspired all the reductionist futilities of the twentieth century, from the obtuse materialism of Marx to the pagan worship of a static material environment, from the Freudian view of the brain as a thermodynamic machine to the zero-sum Malthusian panic over population, treating people more as mouths than as minds." Gilder concluded that Darwinism is "a retrograde retreat to nineteenth-century materialist superstitions, which delude our students that they are learning the facts of science when instead they are imbibing the consolations of a faith-driven materialist myth. In their schools and lives, they deserve some intelligent design."[21]

By "intelligent design," Gilder obviously did not mean a designed economy. According to an essay posted on the Ludwig von Mises Institute blog in 2004, "the core of the argument against intelligent design is that it deserves no credibility because it is neither proven nor provable. However, that argument has an important social parallel. Is it provable that the government, whose only superiority is in the use of coercion, advances Americans' 'general welfare' by its intrusion in every area of life? If not, should we believe in relying on it to make ever more of our choices for us?" The author, Gary Galles, called unwarranted government intrusion "intelligent government design" and argued that "we should surely not teach such unproven ideas in our schools."[22]

This is not what ID theorists mean by "intelligent design." And this is not what George Gilder means by it either. Intelligent design simply means that features of the world that *appear* to be designed may *actually* be designed. It is Darwinism, not ID, that encourages the sort of government interference that conservatives abhor. It is ID, not Darwinism, that encourages openness to what Gilder calls the true source of creativity and wealth, because if design is real then there must be a designer.

Chapter Fifteen

≋≋≋≋≋≋≋≋≋≋≋≋≋

DARWINISM'S WAR ON TRADITIONAL CHRISTIANITY

O n October 21, 2005, Cornell University interim president Hunter R. Rawlings III delivered the annual State of the University address to trustees and faculty. He tackled an "urgent" matter "of great significance to Cornell and to the country as a whole, a matter with fundamental educational, intellectual, and political implications"—namely, "the challenge to science posed by religiously based opposition to evolution, described, in its current form, as 'intelligent design.'"[1]

Rawlings began by quoting the Discovery Institute: "The scientific theory of intelligent design holds that certain features of the universe and of living things are best explained by an intelligent cause, not an undirected process such as natural selection. Note: Intelligent design theory does *not* claim . . . that the intelligent cause must be a 'divine being' or a 'higher power' or an 'all-powerful force.'"

Rawlings went on to argue (despite the disclaimer he had just quoted) that ID "is, at its core, a religious belief." In contrast to ID, Rawlings said, "evolutionary theory says nothing about the existence or non-existence of God." As an example of how best to respond to ID's "assault on science and reason" he cited the 1896 magnum opus of Cornell's co-founder, Andrew Dickson White. White's *History of the Warfare of Science with Theology in Christendom* claimed that after centuries of Christian opposition to science people were finally witnessing "the last expiring convulsions of the old

Guess what?

❦ The three major branches of the Christian tradition share the conviction that human beings were made by design—a conviction Darwinism denies.

❦ Darwinist university professors claim that Christianity should be in cultural zoos and that the pope is "a corpse in a funny hat wearing a dress."

❦ Taxpayers' money is now being used to promote religious denominations that approve of Darwinism.

169

theologic theory" as Darwinism swept away "outworn creeds and noxious dogmas."[2]

Yet historians have known for decades that White's "warfare metaphor" is seriously misleading. Before Darwin, science and theology in Christendom generally got along quite well. Indeed, most of the time they were mutually supportive. Serious conflict erupted only after 1859, and then only because Darwinism declared war on traditional Christianity.[3]

The Christian Tradition

As of 2005, Christianity had far more adherents than any other religion in the world. About 33 percent of the world's six and a half billion people identify themselves as Christians, and half of those are Roman Catholics. About 21 percent of the world identify themselves as Muslims, 12 percent as non-religious, 14 percent as Hindus. Buddhists are 6 percent, Jews .23 percent, other religions 12.6 percent, and atheists 2 percent.[4]

Traditional Christianity is not the same as young-Earth creationism. The three major branches of Christianity—Roman Catholicism, Eastern Orthodoxy, and Protestantism—hold some basic affirmations in common, but belief in a literal six-day creation and 6,000-year-old Earth is not one of them.

Christians are traditionally united not only by the Old and New Testaments, but also by the creedal affirmations that (1) there is "one God the Father Almighty, Maker of heaven and earth, and of all things visible and invisible," (2) for our salvation God became incarnate in Jesus Christ, who is "truly God and truly man," with a human nature consisting of a soul and body "like us in all respects except for sin," and (3) God speaks to us through the Holy Spirit, the "giver of life."[5]

For eighteen centuries, Christian theologians understood these central doctrines to mean that (1) God created everything from nothing, (2) God planned the incarnation—and thus human beings—from the beginning, and (3) God continues to interact with the creation.

In the fourth century A.D. Athanasius emphasized that everything that exists is preceded by "its pattern in God," who "fashioned the race of men" in "His own image." Christ's body, which "is of no different sort from ours," was thus made according to God's eternal plan. In the fifth century, Augustine wrote: "The Wisdom of God, by which all things have been made, contains everything according to design before it is made." In particular, God ordained the form of human beings "before the ages."[6]

In the seventh century, Maximus the Confessor (regarded as a leading theologian by Eastern Orthodoxy) wrote that God's original purpose for creating the universe was to make human beings microcosms that unite the mental, spiritual, and physical aspects of creation with each other and with Him. In the thirteenth century, Thomas Aquinas (the leading theologian of Roman Catholicism) wrote that created beings are preceded by "ideas in the divine mind" just as "the likeness of a house pre-exists in the mind of the builder." In particular, the human soul is "made to the image of God," and "God fashioned the human body" in order to "make it suitably proportioned to the soul."[7]

During the Protestant Reformation of the sixteenth century, Martin Luther emphasized that "the creation is not fortuitous but the exclusive work of divine foresight." Above all, man and woman were "created by the special plan and providence of God." According to John Calvin, the "most excellent example" of all God's works is humankind. The "chief seat" of God's image is in the human mind and heart, though some "scintillations" shine in every part, including the body.[8]

So in traditional Christian doctrine, the world was created and designed by God,

A Book You're Not Supposed to Read

Total Truth: Liberating Christianity from Its Cultural Captivity, by Nancy Pearcey; Wheaton, IL: Crossway Books, 2004.

and human beings were planned from the beginning. A great many of the world's people believe that these statements are true. For our present purpose, though, what matters is not whether they are true, but what Darwinism has to say about them.

Darwinism vs. Christianity

Charles Darwin acknowledged "the extreme difficulty or rather impossibility of conceiving this immense and wonderful universe" as "the result of blind chance or necessity," and he wrote that life had "been originally breathed by the Creator into a few forms or into one." According to Darwin, however, all living things after the first have descended with modification from a common ancestor by natural selection acting on small variations. But "natural selection means only the preservation of variations which independently arise," while "no shadow of reason can be assigned for the belief that variations . . . were intentionally and specially guided."[9]

Darwin concluded: "There seems to be no more design in the variability of organic beings, and in the action of natural selection, than in the course which the winds blow." Although "I cannot look at the universe as the result of blind chance," he wrote, "yet I can see no evidence of beneficent design, or indeed of design of any kind, in the details."[10]

So in Darwin's thinking a deity may have designed the universe and its laws, but human beings are unplanned and undesigned. In his influential book *The Meaning of Evolution*, Darwinian paleontologist George Gaylord Simpson wrote: "Man is the result of a purposeless and natural process that did not have him in mind. He was not planned." In 1970, molecular biologist Jacques Monod said that because "the mechanism of Darwinism is at last securely founded . . . man has to understand that he is a mere accident." And in 1977 Stephen Jay Gould wrote in *Ever Since Darwin*: "Biology took away our status as paragons created in the image of God."[11]

Many Darwinists are virulently anti-Christian. Richard Dawkins once said religion "is one of the world's great evils, comparable to the smallpox virus but harder to eradicate," and Daniel C. Dennett thinks Christianity should be "preserved in cultural zoos." A common sight in American college towns is cars displaying Darwin fish—deliberate mockeries of the traditional symbol of Christianity, with feet underneath and "Darwin" inside. One variant has a Darwin fish raping a Christian fish.[12]

Anti-Christian zealots are often in the forefront of attacks on intelligent design. In 2005, the chairman of the University of Kansas Religious Studies Department, atheist Paul Mirecki, proposed to teach a course titled "Intelligent Design Creationism and Other Mythologies." Mirecki boasted on a web site that "fundies" would see the course as a "slap in their big fat face." He also endorsed a description of Pope John Paul II as "a corpse in a funny hat wearing a dress."[13]

Even disregarding such excesses, it is clear that there is a fundamental conflict here. It is not between religion and science, or even between Christianity and evolution, but between traditional Christianity and Darwinism. Although the latter may allow for the existence of a deity, it is not the God of traditional Christianity, who created human beings in His image. The contradiction couldn't be sharper, and most attempts to blunt it end up by abandoning traditional Christianity.

Surrendering on Darwin's Terms

Florida State University philosopher Michael Ruse (who is not a Christian) considers Darwinism so well established that Christians should accept it as a fact. Although this means that "there is simply no guarantee

Darwinist considers religion worse than smallpox

"Faith is one of the world's great evils, comparable to the smallpox virus but harder to eradicate."

—Darwinist **Richard Dawkins**, 1997

that humans or anything else would have evolved," Ruse assures Christians: "It is still open to you to accept that God did the job. More likely, if you accept God already, it is still very much open to you to think of God as great inasmuch as He has created this wonderful world." In other words, a Darwinian who really, *really* wants to be a Christian can be a "Christian" of sorts—just not a traditional one.[14]

Like Ruse, Roman Catholic biologist Kenneth R. Miller considers Darwinism well established; indeed, "the intellectual triumph of Darwin's great idea is total." Miller argues that the inherent unpredictability of evolution was essential to God's plan to create human beings with free will. "If events in the material world were strictly determined," he writes, "then evolution would indeed move toward the predictable outcomes that so many people seem to want.... As material beings, our actions and even our thoughts would be preordained, and our freedom to act and choose would disappear." Miller concludes by declaring that he believes "in Darwin's God." In the Christian tradition, however, human freedom is an attribute of our non-material souls rather than a product of material evolution. Darwin's God is not the God of traditional Christianity.[15]

Stephen Jay Gould wrote in 1997 that there should be no conflict between science and religion "because each subject has a legitimate magisterium, or domain of teaching authority—and these magisteria do not overlap (the principle that I would like to designate as NOMA, or 'nonoverlapping magisteria'). The net of science covers the empirical universe.... The net of religion extends over questions of moral meaning and value." In other words, the world of objective reality belongs to science (and

A Darwinist's opinion of Pope John Paul II

"A corpse in a funny hat wearing a dress."

Statement endorsed by Kansas professor **Paul Mirecki**, 2005 thus, for Gould, to Darwinism), while religion is confined to the realm of subjective feelings and imagination.[16]

But traditional Christians cannot agree to Gould's terms of surrender. As University of Notre Dame philosopher Alvin Plantinga put it in 1997, Darwinists claim "that human beings are, in an important way, merely accidental; there wasn't any plan, any foresight, any mind, any mind's eye involved in their coming into being. But of course no Christian theist could take that seriously for a minute."[17]

Polls taken over the past few years have consistently shown that a large majority of the American people not only describe themselves as Christian, but also reject Darwinism. Since Darwinists depend for their livelihood largely on tax money, and since their attempts at accommodation have largely failed, they try to win people over in the United States by calling Darwinism simply "evolution." Eugenie C. Scott of the National Center for Science Education (NCSE) recommends starting by defining the issue as "change through time." The rest comes later.[18]

Using this approach, Darwinists have convinced many Christian clergy to support their cause, and to good effect. Scott wrote in 2002: "I have found that the most effective allies for evolution are people of the faith community. One clergyman with a backward collar is worth two biologists at a school board meeting any day!"[19]

In 2004, evolutionary biologist Michael Zimmerman, dean of the College of Letters and Sciences at the University of Wisconsin–Oshkosh, started collecting signatures from clergy who support evolution. By 2006, Zimmerman had persuaded over ten thousand clergy to sign a letter "affirming the teaching of the theory of evolution as a core component of human knowledge" and asking that science and religion remain "two very different, but complementary, forms of truth." On "Evolution Sunday," February 12, 2006 (the 197th anniversary of Charles Darwin's birth), several hundred churches nationwide devoted their Sunday services to praising Darwinism.[20]

Meanwhile, not all Christians were persuaded. Hundreds of thousands of more traditional Christians were buying souvenir cards in Rome with the words of Pope Benedict XVI: "We are not some casual and meaningless product of evolution. Each of us is the result of a thought of God."[21]

Roman Catholicism

In 1950, Pope Pius XII's encyclical *Humani Generis* affirmed that human souls are created by God but permitted research concerning the evolutionary doctrine that the human body comes "from pre-existent and living matter"—as long as "the reasons for both opinions, that is, those favorable and those unfavorable to evolution, be weighed and judged." The pope warned against assuming that "the origin of the human body from pre-existing and living matter were already completely certain."[22] Almost half a century later, in a message to the Pontifical Academy of Sciences, Pope John Paul II said that science had progressed to the point where "some new findings lead us toward the recognition of evolution as more than an hypothesis." Nevertheless, the pope continued: "Rather than speaking about the theory of evolution, it is more accurate to speak of the theories of evolution. The use of the plural is required here—in part because of the diversity of explanations regarding the mechanism of evolution, and in part because of the diversity of philosophies involved. There are materialist and reductionist theories, as well as spiritualist theories. Here the final judgment is within the competence of philosophy and, beyond that, of theology."[23]

Darwinists have widely advertised Pope John Paul II's statement as a blanket endorsement of their position by the Roman Catholic Church, even though the pope made a point of *not* endorsing their claim that unguided natural mechanisms are sufficient to explain evolution. In 2004, the Vatican's International Theological Commission issued a statement acknowledging general agreement among scientists "that the first organism dwelt on this planet about 3.5 to 4 billion years ago. Since it has been demonstrated that all living organisms on earth are genetically related, it is virtually certain that all living organisms have descended from this first organism." But, "the message of Pope John Paul II cannot be read as a blanket approbation of all theories of evolution, including those of a neo-Darwinian provenance which explicitly deny to divine providence any truly causal role in the development of life in the universe."[24]

So the commission endorsed the idea of universal common ancestry (as it applies to the human body) but not theories such as Darwinism that deny design. The statement was important not only because it contradicted the Darwinists' version of Pope John Paul II's 1996 message, but also because the commission's president was Joseph Cardinal Ratzinger, who a year later became Pope Benedict XVI. In his inaugural homily on Easter Sunday, April 22, 2005, the new pope delivered the message now printed on the souvenir cards being sold near the Vatican.[25]

In July 2005, the pope's longtime friend Christoph Cardinal Schönborn of Vienna published an essay in the *New York Times* criticizing Darwinists who "sought to portray our new pope, Benedict XVI, as a satisfied evolutionist." The cardinal reaffirmed "the perennial teaching of the Catholic Church about the reality of design in nature," which "the human intellect can readily and clearly discern." He concluded: "Faced with scientific claims like neo-Darwinism . . . invented to avoid the overwhelming evidence for purpose and design found in modern science, the Catholic Church will again defend human reason by proclaiming that the immanent design evident in nature is real."[26]

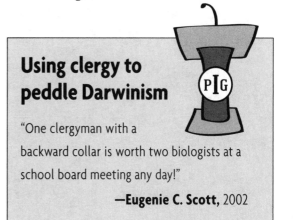

Using clergy to peddle Darwinism

"One clergyman with a backward collar is worth two biologists at a school board meeting any day!"

—Eugenie C. Scott, 2002

But Jesuit priest George Coyne, director of the Vatican Observatory, announced that Cardinal Schönborn (and by implication the pope) was wrong. According to Coyne: "There appears to exist a nagging fear in the Church that a universe, which science has established as evolving... through a process of random genetic mutations and natural selection, escapes God's dominion. That fear is groundless. Science is completely neutral with respect to philosophical or theological implications." Coyne concluded: "It is difficult to believe that God is omnipotent and omniscient in the sense of many of the scholastic philosophers. For the believer, science tells of a God who must be very different from God as seen by them."[27]

This logic-challenged priest—science is theologically neutral yet leads to a *different* God—has the arrogance to lecture a pope and a cardinal on Catholic doctrine. Making matters worse, American Darwinists portray him as a Vatican spokesman![28]

An Establishment of Religion

The pope on Darwinism

"We are not some casual and meaningless product of evolution. Each of us is the result of a thought of God."

—Pope Benedict XVI,
inaugural homily, 2005

The First Amendment to the U.S. Constitution begins: "Congress shall make no law respecting an establishment of religion," a prohibition that has since been extended to all state and local governments. Although the history of its interpretation is long and complicated, the First Amendment clearly prohibits the government from favoring the views of one religious group over another.

State universities are government institutions and the National Science Foundation (NSF) is a federal agency. In

2001, the NSF granted more than $450,000 to the University of California–Berkeley to develop a pro-Darwin website in cooperation with the NCSE. The website, "Understanding Evolution," now includes the following statement: "The misconception that one always has to choose between science and religion is incorrect. Of course, some religious beliefs explicitly contradict science (e.g., the belief that the world and all life on it were created in six literal days); however, most religious groups have no conflict with the theory of evolution or other scientific findings. In fact, many religious people, including theologians, feel that a deeper understanding of nature actually enriches their faith."[29]

The website also includes a link to the NCSE's "Voices for Evolution," a collection of statements from more than a dozen religious organizations affirming the compatibility of evolution with their teachings. In one example, the Episcopal bishop of Atlanta wrote: "Evolution as a contemporary description of the how of creation is anticipated in its newness by the very fluidity of the biblical text." And the General Assembly of the Presbyterian Church (USA) declared: "There is no contradiction between an evolutionary theory of human origins and the doctrine of God as Creator."[30]

Political scientist John G. West wrote in 2004: "Taxpayers might wonder why it's the government's business to tell them what their religious beliefs about evolution should or shouldn't be." And legal scholar Francis J. Beckwith pointed out that since the government "may not aid, foster, or promote one religion or religious theory against another," the NSF and University of California "should not be in the business of directly funding the propagation of one view as acceptable theological opinion on any matter."[31]

In 2005, a California resident sued the NSF and University of California for violating the First Amendment, but district court judge Phyllis J. Hamilton dismissed the lawsuit. So the governments of the United States and California now officially endorse religious views—and *only* those religious views —that are acceptable to Darwinists.[32]

What's next?

Chapter Sixteen

❧❧❧❧❧❧❧❧❧❧❧❧❧

AMERICAN LYSENKOISM

I n fall 2005, I visited a biology graduate student at a major American university. He had been helping behind-the-scenes to form an undergraduate club on his campus that would provide a forum for discussions about intelligent design. He feared that if his role became known it would mean the end of his career, so he communicated with me using a pseudonym, an off-campus e-mail account, and a cell phone.

He could not risk even being seen with me. Driving a rental car, I picked him up at the appointed time on a street corner several blocks from campus, and we drove to lunch at an out-of-the-way restaurant in another town. Afterwards, I drove him back to another off-campus location.

Were we just being paranoid?

The year before, intelligent design theorist William Dembski had written: "Doubting Darwinian orthodoxy is comparable to opposing the party line of a Stalinist regime. What would you do if you were in Stalin's Russia and wanted to argue that Lysenko was wrong? . . . That's the situation we're in."[1]

Trofim D. Lysenko (1898–1976) was a Ukrainian agronomist who rose to a high position in the Soviet scientific establishment under the dictatorship of Joseph Stalin. In the 1930s and 1940s, he promoted various ill-founded ideas about heredity and used his political power ruthlessly to

Guess what?

- When the Soviet government supported one side in a scientific controversy eighty years ago, the result was Lysenkoism, which persecuted scientists and obstructed progress in biology.
- Lysenkoism is now rearing its ugly head in the U.S. as Darwinists use their government positions to destroy the careers of their critics.
- A federal judge has declared that requiring critical analysis of Darwinism in public schools is unconstitutional.

suppress his critics, many of whom lost their jobs and some of whom suffered imprisonment or even death.

Was Dembski justified in comparing American critics of Darwinism to Soviet critics of Lysenkoism? According to some Darwinists, it's the other way around.

The Darwinists' View

Retired physicist Mark Perakh, who grew up in the former Soviet Union, writes: "The anti-Lysenkoist stand of the ID advocates is...ludicrous given the similarity of their denial of Darwinian biology to the denial of the neo-Darwinian synthesis by the Lysenkoists." Perakh continues: "From my experience both with Marxism and with the realities of the Soviet system, I can assert that...it is ID advocates whose behavior is reminiscent of the oppressive Soviet regime" since they subject Darwinists "to continuous denunciations, verbal assaults, derision, and ultimately to dismissal from their positions."[2]

According to philosopher Robert T. Pennock: "In the former Soviet Union, Darwinian evolution was rejected on ideological grounds.

Who won the Cold War?

"Doubting Darwinian orthodoxy is comparable to opposing the party line of a Stalinist regime. What would you do if you were in Stalin's Russia and wanted to argue that Lysenko was wrong? You might point to paradoxes and tensions in Lysenko's theory of genetics, but you could not say that Lysenko was fundamentally wrong or offer an alternative that clearly contradicted Lysenko. That's the situation we're in."

—William A. Dembski, 2004

Because the Communist Party denounced the Darwinian view in favor of Lysenkoism, a variant of Lamarckism that was more in line with Party ideology, biological research was set back for a generation. ID-ology could have the same effect in this country, if it succeeds in its lobbying efforts." In his 2005 college textbook *Evolution*, Douglas J. Futuyma writes that under Lysenko "Mendelism and Darwinism came to be viewed as unwelcome or dangerous," and "the result was a disaster for Soviet food production and the Soviet people." Futuyma concludes that intelligent design poses a similar threat today.[3]

Former assistant vice chancellor of the University of California–San Francisco Christopher Scott writes: "Ideology masquerading as scientific legitimacy is sweeping into American classrooms" in the form of opposition to Darwinism. "Government pressure on educators," Scott continues, is "a disturbing American vision. But it's not unique to America. In the '20s, scientists worldwide vigorously debated the mechanisms of evolution. On one side were the Darwinians; on the other were supporters of Jean-Baptiste Lamarck." Lamarck's ideas were "very handy for Joseph Stalin, who rejected any doctrine—like Darwinism—that challenged socialism." Scott warns that "Stalin's political solution choked off scientific progress."[4]

Soviet Lysenkoism

As is always the case with complex phenomena, historians differ in their interpretations of Lysenkoism, but several relevant facts stand out.

First, from the very beginning Darwinism was central to the Soviet worldview, largely because of its historical approach to human origins and its materialistic rejection of religion. The claims by Robert Pennock and Christopher Scott that Soviet Communists and Stalin rejected Darwinism are false.[5]

Second, Darwin entertained various ideas about heredity, including the Lamarckian inheritance of acquired characteristics. "I think there can be no doubt," he wrote in the sixth edition of *The Origin of Species*, "that use in our domestic animals has strengthened and enlarged certain parts, and disuse diminished them; and that such modifications are inherited." So there were similarities between some of Darwin's views and Lamarck's, though Darwin placed greater emphasis on the role of natural selection than on acquired characteristics.[6]

Third, before the 1920s many scientists—including Gregor Mendel—regarded Mendelian genetics as incompatible with Darwinian evolution. Genes that remain unchanged in the process of descent seemed incapable of providing the modifications required by Darwin's theory. It was largely for this reason that Mendel's work—published in 1866—was completely ignored until 1900 and discounted by many biologists for decades thereafter.[7]

Fourth, in the 1930s even some Western biologists remained skeptical of Mendelian genetics. Genes were unobserved abstractions, there were some well-known exceptions to Mendel's laws, and the role of DNA in heredity had not yet been discovered.[8]

Fifth, science in the Soviet Union was government-supported on a scale unprecedented in history. Driven by a desire to surpass the West, the Soviet Union devoted a larger share of its budget to science than any other industrialized nation. Unfortunately, unprecedented government support also meant unprecedented government entanglement.[9]

Enter Trofim Lysenko. Lysenko seems to have been an opportunist rather than an ideologue. He was not a Marxist, and he never joined the Communist Party. In the late 1920s, he attracted the attention of agricultural scientists by publishing and lecturing about the effects of temperature on plant development. When Stalin's forced collectivization of farms produced widespread famine, the Soviet government asked the scientific establishment to come up with immediate solutions, and Lysenko jumped at the chance. He proposed a physiological treatment he called "vernalization," which he claimed would increase productivity by making it possible to sow winter grain in the spring. Vernalization failed to live up to its promise, and Lysenko went on to propose other solutions. They failed too, but Lysenko succeeded in gaining the support of government officials—including Stalin—who were impatient with the apparent inability of academic biologists to come up with quick solutions to the nation's pressing problems.[10]

Lysenko had little formal training in biology, much less Mendelian genetics. He also initially denied any kinship between his ideas and Lamarckism, but after Isaak Prezent, president of the Society of Marxist Biologists, introduced him to evolutionary theory he eagerly embraced Darwinism—along with its Lamarckian elements. By the 1930s, Lysenko was conflating the physiological process of vernalization with the Lamarckian inheritance of acquired characteristics. When Mendelian biologists criticized him, he simply evaded their arguments and declared that Mendelian genetics was unacceptable because it contradicted Darwinian evolution.[11]

A Book You're Not Supposed to Read

The Lysenko Effect: The Politics of Science, by Nils Roll-Hansen; Amherst, NY: Humanity Books, 2005.

By then, many Western biologists were accepting the "modern synthesis" of Darwinian evolution and Mendelian genetics, but Soviet minister of agriculture Jakov Jakovlev supported Lysenko by declaring Mendelism to be incompatible with true Darwinism. In 1937, Prezent praised Lysenko for "marching...under the banner of reconstruction of biological science on the basis of Darwinism raised to the level of Marxism," while he demonized the Mendelians as "powers of darkness."[12]

If government officials and Darwinist ideologues had not come to Lysenko's rescue, the Mendelians would probably have prevailed—as they did outside the Soviet Union—because they had better science on their side. Lysenko's Stalinist suppression of Mendelians in the 1940s made matters much worse, but the underlying problem was that the government-supported scientific establishment had chosen to support one side in a scientific dispute. For many years, biologists in the Soviet Union were persecuted *by the government* if they challenged the official view of Darwinian orthodoxy or defended Mendelian genetics.[13]

So, contrary to the claims of Pennock, Futuyma, and Christopher Scott, the scientific conflict underlying Lysenkoism was not Lamarckism against Darwinism, but classical Darwinism (which had undeniably Lamarckian elements) against the new Mendelian genetics. The present conflict between neo-Darwinism and intelligent design resembles Lysenkoism in the sense that the Darwinists are still opposing new ideas.

But what about Perakh's accusation that ID theorists resemble Lysenkoists in the way they subject their critics to "denunciations, verbal assaults, derision, and ultimately to dismissal from their positions?"

Are those English hooligans? No, they're Darwinists

"The only appropriate response should involve some form of righteous fury, much butt-kicking, and the public firing of some teachers, many schoolboard members, and vast numbers of sleazy far-right politicians...It's time for scientists to break out the steel-toed boots and brass knuckles, and get out there and hammer on the lunatics and idiots."

—University of
Minnesota professor
Paul Z. Myers

Denunciation and Dismissal?

To support his accusation, Perakh cites four examples. In the first two, William Dembski refers to the arguments of his critics in disparaging terms, and in the third Dembski objects that a critic whose highest academic degree is an M.A. in geography is unqualified to write as an authority on the bacterial flagellum. These examples are typical of academic disputes; not one of them rises even to the level of derision.

Perakh's only other example is a comment I made after debating Darwinists before the Ohio State Board of Education in 2002 that I felt as though I "needed a shower." Perakh calls this an example of the "rude personal attacks" that are "a staple in the pro-ID literature." What Perakh neglected to mention was that I made the comment only after one of the Darwinists in the debate had begun with a series of personal attacks on me. The verbal assaults had come from the Darwinist.[14]

In March 2005, the National Science Teachers Association conducted an informal poll of its 55,000 members to find out if teachers were feeling "pressured to include creationism, intelligent design, or other nonscientific alternatives to evolution in their science classroom," or "pushed to de-emphasize or omit evolution" from their curriculum. Less than a third of the 1,050 teachers who responded indicated that they had received any such pressure, and the "pressure" (which was not further specified) had come predominantly from students and parents, not school administrators. Nothing in the poll suggested that the teachers had been subjected to derision or denunciation, much less threat of dismissal.[15]

University of Kansas religious studies professor Paul Mirecki proposed in 2005 to teach a course provocatively titled "Intelligent Design Creationism and Other Mythologies." Mirecki's proposed course was withdrawn, but he subsequently claimed to have suffered a pre-dawn beating on a deserted country road by two men who (he said) made references to the ID-as-mythology course. Skeptics question Mirecki's story, which is still under investigation by the county sheriff, but if it's true, his assailants deserve to be prosecuted.[16]

Yet even if Mirecki's as-yet-unsubstantiated assault story turns out to be true, it is not an example of Lysenkoism. What made Lysenkoism so destructive was its backing by the coercive power of the government. There has been no allegation (much less evidence) that Mirecki's beating was instigated or condoned by the government.

Actually, there are lots of examples of government-backed denunciation and dismissal in the present controversy, but all of them involve Darwinists persecuting their critics. What follows is just a small sample.

Will the Real Lysenkoists Please Stand Up?

Brian R. Leiter is a philosopher and lawyer who teaches in the School of Law at the University of Texas–Austin, a public institution. Leiter is also an outspoken Darwinist and atheist. In 2003, Baylor University legal scholar Francis J. Beckwith published a book arguing that the U.S. Constitution does not bar intelligent design from public school classrooms. In 2004, Harvard Law School student Lawrence VanDyke published a favorable review of Beckwith's book in the *Harvard Law Review*, and Leiter accused him of "scholarly fraud"—largely because (according to Leiter) the student ignored the "massing evidence" that "established evolution's truth beyond reasonable doubt." Leiter didn't stop with denunciation; he went on to warn VanDyke against "entering law teaching."

On the tube

From *Friends*: Ross on evolution

Ross: You don't believe in evolution?

Phoebe: I don't know, it's just, you know…monkeys, Darwin, you know, it's a, it's a nice story, I just think it's a little too easy.

Ross: Too easy? Too…the process of every living thing on this planet evolving over millions of years from single-celled organisms is…is too easy?

Phoebe: Yeah, I just don't buy it.

Ross: Uh, excuse me. Evolution is not for you to buy, Phoebe. Evolution is scientific fact, like, like, like the air we breathe, like gravity.

Phoebe: Oh, okay, don't get me started on gravity.

Ross: You, uh, you don't believe in gravity?

Phoebe: Well, it's not so much that, you know, like, I don't believe in it, you know, it's just…I don't know, lately I get the feeling that I'm not so much being pulled down as I am being pushed.

Beckwith was later quoted in *National Review* as saying: "Leiter's apparent intention to employ his own celebrity and academic stature to crush a young man's spirit and his future job prospects in the legal academy, and to do so by means of blacklisting and mean-spirited McCarthyesque intimidation tactics, is absolutely unjustified."[17]

Leiter may or may not destroy VanDyke's career, but some Darwinist professors at Ohio State University (OSU)—a public institution—are now trying to destroy another student's career by preventing him from getting his doctorate. Bryan Leonard is a high school science teacher who for the past few years has been working on a Ph.D. in science education. His dissertation research focused on these questions: When students are taught the scientific data both supporting and challenging macroevolution, do they maintain or change their beliefs over time? What empirical, cognitive, and/or social factors influence students' beliefs? Leonard finished his research and wrote his dissertation, and his defense was scheduled for June 2005.

Central Washington University biology professor Steven Verhey published similar research involving his students in the journal *BioScience* in 2005. Darwinists didn't object, because Verhey's goal was to convert students to Darwinism, but Leonard had angered Darwinists in 2004 by helping to write the model lesson plan for the Ohio curriculum that encourages critical study of evolution. When Leonard recommended a similar approach in testimony before the Kansas State Board of Education in May 2005 (Figure 8), the Darwinists retaliated.

Although Leonard had gone through normal procedures and received proper approval to conduct his research, OSU professors Brian McEnnis, Steve Rissing, and Jeffrey McKee accused Leonard of "unethical" conduct, primarily on the grounds that his research was predicated on "a fundamental flaw: there are no valid scientific data challenging macroevolution." So Leonard's research (they claimed) involved "deliberate miseducation of these students, a practice we regard as unethical."

The OSU Darwinists then invoked some procedural technicalities—widely ignored in the case of other Ph.D. candidates—to demand that Leonard's dissertation defense be postponed. McKee subsequently compared two biologists who were members of Leonard's dissertation committee to "parasitic ticks hiding in the university's scalp." McKee wrote that he had learned as a boy "to twist the ticks when taking them out, so their heads don't get embedded in the skin. Others prefer burning them off. What fate awaits OSU's ticks remains to be seen."[18]

Government-employed Darwinists don't stop at denouncing their critics and threatening future careers; they also threaten and destroy the careers of qualified teachers and scientists who challenge Darwinian orthodoxy. Paul Z. Myers is a biology professor at the University of Minnesota–Morris, a public institution. Myers recommends "the public firing and humiliation" of teachers who dare to speak approvingly about intelligent design. He goes even further: "I say, screw the polite words and careful rhetoric. It's time for scientists to break out the steel-toed boots and brass knuckles."[19]

The public firings recommended by Myers have already begun. The 2001 dismissal of Roger DeHart from his public high school biology position was just the tip of a huge iceberg. In 2003, Dr. Nancy Bryson was head of the Division of Science and Mathematics at the Mississippi University for Women, a public institution. After she presented an honors forum titled "Critical Thinking on Evolution" that included scientific criticisms of chemical and biological evolution, a senior biology professor read to the audience a previously prepared statement calling her presentation "religion masquerading as science" and accusing her of being unqualified to talk about evolution. The next day she was informed that her contract as division head would not be renewed and that she would probably not be retained as a faculty member. She subsequently had to find work elsewhere.[20]

Dr. Caroline Crocker is a biologist with a Ph.D. in immunopharmacology. In January 2003 she became a visiting professor at George Mason

University, a public institution just outside Washington, D.C. While covering the section on evolution in a 2004 cell biology course, she gave one lecture on evidentiary problems with Darwin's theory and briefly mentioned the controversy over intelligent design. At the end of the lecture she said to students, "Well, you need to make up your own mind. Think about it for yourself." One student complained verbally to Crocker's supervisor that she was teaching creationism, though another student testified in writing that this was false. Nevertheless, the supervisor told Crocker that as a disciplinary measure she would be barred from lecturing and restricted to teaching lab sections in the spring. Crocker's contract was not renewed.[21]

The Darwinists who drove Crocker from her job—like Darwinist Barbara Forrest at Southeastern Louisiana University; the Darwinists who harassed Richard Sternberg at the Smithsonian; president Timothy White of the University of Idaho; Hector Avalos at Iowa State University; the school administrators who drove out Roger DeHart; Brian Leiter at the University of Texas; Paul Z. Myers at the University of Minnesota; Brian McEnnis, Steve Rissing, and Jeffrey McKee at Ohio State University; and the Mississippi University for Women Darwinists who dismissed Nancy Bryson—are all public employees. These and other Darwinists are using

Judgment day

"We will enter an order permanently enjoining Defendants [the Dover school board] from . . . requiring teachers to denigrate or disparage the scientific theory of evolution."

—Judge John E. Jones III, 2005

their public positions—*government* positions—to enforce political correctness by threatening students' careers, dismissing credentialed teachers, and harassing qualified scientists who dare to question the Darwinist party line. The fact that such people pretend to respect academic freedom makes them hypocrites; the fact that they are employed by the government makes them Lysenkoists.

Outlawing Criticisms of Darwinism

In December 2005, American Lysenkoism staged its biggest coup. U.S. district court judge John E. Jones III permanently prohibited the Dover, Pennsylvania, school district not only from teaching intelligent design, but also "from requiring teachers to denigrate or disparage the scientific theory of evolution." By judicial fiat, Darwinism is now State Science, at least in central Pennsylvania. All criticism of it in the public schools is discouraged under penalty of law. Lysenko would have been delighted.[22]

The United States, thank God, is very different from the former Soviet Union. Despite the hate-America rhetoric spewed by some leftists, the FBI is not the KGB and George W. Bush is not Joseph Stalin. Indeed, President Bush opposes the suppression of either Darwinism or intelligent

Fear factor

"I lost my job at George Mason University for teaching the problems with evolution. Lots of scientists question evolution, but they would lose their jobs if they spoke out."

—**Caroline Crocker**, quoted in the
Washington Post, February 5, 2006

design and encourages free and open discussion of both. The breeding ground of American Lysenkoism is not the White House, but in publicly supported institutions that have been taken over by ruthless Darwinists and courtrooms that have been taken over by judicial megalomaniacs.

The biology graduate student I spirited off to lunch in fall 2005 was not paranoid. He had good reason to fear for his career. Except for the obvious differences between our constitutional democracy and Stalin's brutal dictatorship, Dembski's comparison was justified: a critic of Darwinism in America today is in a position comparable to a critic of Lysenkoism in the former Soviet Union.

❧❧❧❧❧❧❧❧❧❧❧❧❧❧

Chapter Seventeen

SCIENTIFIC REVOLUTION

As a college geology student in the early 1960s, I breathed the air of scientific revolution. Five decades earlier, Alfred Wegener had suggested that the continents were slowly drifting apart from what used to be a single land mass. Wegener's theory of continental drift was still controversial, but several of my professors defended it and their graduate students were doing research that eventually led to its widespread acceptance.[1]

In the midst of the excitement, I read Thomas Kuhn's now-famous book *The Structure of Scientific Revolutions*. Although some of Kuhn's more radical conclusions are very questionable, his description of what happens in a scientific revolution helped me to put the controversy over continental drift in a historical perspective. It can also help us understand the controversy over Darwinism and intelligent design—and show us how it is likely to turn out.

The Structure of Scientific Revolutions

According to Kuhn, the activities of scientists fall into two different categories. First, they spend almost all their time doing what he called "normal science." This is "research firmly based upon one or more past scientific achievements...that some particular scientific community

Guess what?

- ❧ The controversy between Darwinism and intelligent design has the characteristics of major scientific revolutions in the past.
- ❧ Darwinists are losing power because they treat with contempt the very people on whom they depend the most: American taxpayers.
- ❧ The outcome of this scientific revolution will be decided by young people who have the courage to question dogmatism and follow the evidence wherever it leads.

acknowledges for a time as supplying the foundation for its further practice." Kuhn called such achievements "paradigms." Normal science articulates an existing paradigm, determines facts, and fits them to the paradigm. It is not engaged in seeking new ways to understand the world, but in solving puzzles.[2]

"No part of the aim of normal science is to call forth new sorts of phenomena," Kuhn wrote. "Nor do scientists normally aim to invent new theories, and they are often intolerant of those invented by others." Nevertheless, new theories do appear and paradigms do change. These "scientific revolutions," Kuhn wrote, "are inaugurated by a growing sense . . . often restricted to a narrow subdivision of the scientific community, that an existing paradigm has ceased to function adequately" in exploring nature. Even then, revolution "occurs only after the sense of crisis has evoked an alternative candidate for paradigm."[3]

How do scientists test competing paradigms? At one point Kuhn wrote: "The decision to reject one paradigm is always simultaneously the decision to accept another, and the judgment leading to the decision involves the comparison of both paradigms with nature *and* with each other." This sounds like the "inference to the best explanation" we encountered in previous chapters.[4]

In most of his book, however, Kuhn suggested that paradigms compete independently of the evidence from nature: "When paradigms enter, as they must, into a debate about paradigm choice, their role is necessarily circular. Each group uses its own paradigm to argue in that paradigm's defense." This became Kuhn's legacy and the source of most of the controversy surrounding his work. If paradigms are essen-

Out with the old

"A new scientific truth does not triumph by convincing its opponents and making them see the light, but rather because its opponents eventually die, and a new generation grows up that is familiar with it."

—Quantum physicist **Max Planck**, 1949

tially circular (Kuhn called them "incommensurable"), then the choice between them depends not on logic and evidence but on irrational psychology and power politics. A new paradigm succeeds not because it provides a better description of nature, but because it defeats the old paradigm in a struggle for survival.[5]

Kuhn saw a "very nearly perfect" analogy between scientific revolutions and Darwinian evolution: "The resolution of revolutions is the selection by conflict within the scientific community of the fittest way to practice future science. The net result of a sequence of such revolutionary selections, separated by periods of normal research, is the wonderfully adapted set of instruments we call modern scientific knowledge."[6]

Kuhn called this "scientific progress," but his critics called it relativism. No theory is any truer than any other—just more successful in the struggle for existence. Yet some theories really *are* truer than others; if they weren't, bridges would collapse, airplanes would crash on takeoff, agriculture would fail, and patients would die on the operating table. So Kuhn's relativism fails to explain scientific progress.[7]

Nevertheless, he provided some interesting descriptions of what happens during scientific revolutions. In particular, he identified four characteristics that are relevant here. First, scientific revolutions are often marked by disputes over the definition of science. Kuhn wrote: "The reception of a new paradigm often necessitates a redefinition of the corresponding science," the "standard that distinguishes a real scientific solution from a mere metaphysical speculation." For example: "Newton's theory of gravity was resisted because "gravity, interpreted as an innate attraction between every pair of particles of matter, was an occult quality" like the medieval "tendency to fall." Critics of Newtonianism claimed that it was not science and "its reliance upon innate forces would return science to the Dark Ages."[8]

Second, like a political revolution, a scientific revolution typically divides society "into competing camps or parties, one seeking to defend the old institutional constellation, the others seeking to institute some new ones." The camp defending the old paradigm uses every means at its disposal, including all of its professional societies and publications, to resist the challenger. Since careers are at stake, things can get personal. Eventually, both parties "resort to the techniques of mass persuasion."[9]

Third, the real issue at stake in a scientific revolution is "which paradigm should in the future guide research. . . . A decision between alternate ways of practicing science is called for, and in the circumstances that decision must be based less on past achievement than on future promise."[10]

And fourth, scientific revolutions are typically precipitated by people who "have been either very young or very new to the field whose paradigm they change," because they are committed "less deeply than most of their contemporaries to the world view and rules determined by the old paradigm." According to quantum physicist Max Planck: "A new scientific truth does not triumph by convincing its opponents and making them see the light, but rather because its opponents eventually die, and a new generation grows up that is familiar with it."[11]

Chapter Twelve examined disputes over the definition of science; this chapter will focus on the other three characteristics described above, all of which suggest that we are in the midst of a major scientific revolution that Darwinism will lose and intelligent design will win.

Why Darwinism Will Lose

Darwinism will lose, most importantly, because of the evidence. Even though Darwinists have had almost 150 years to find some, the evidence for their view is underwhelming, at best. Otherwise, we wouldn't be reading almost every month about some discovery or other that finally "proves" it.

On the other hand, in only a tenth of the time ID theorists have come up with at least preliminary evidence for their case. Otherwise, Darwinists wouldn't be trying so hard to disprove it, ridicule it, or exclude it from serious discussion. If there weren't any evidence for design, *Nature* wouldn't print such nonsense as "even if all the data point to an intelligent designer, such an hypothesis is excluded from science because it is not naturalistic."[12]

Since the Darwinists are losing ground on the evidence, they resort more and more to censorship. "There's no debate!" shouts the Darwin-only crowd. But everyone who's been paying attention knows that there *is* a debate. And anyone who studies American history knows that telling people they are not allowed to talk about something is the tactic *least* likely to succeed in the land of the free and the home of the brave.

Teeth marks from a Darwinist

"Ewww…intelligent design people! They're just buck-toothed, Bible-pushing nincompoops with community-college degrees who're trying to sell a gussied-up creationism to a cretinous public! No need to address their concerns or respond to their arguments. They are Not Science. They are poopy-heads. There. I just saved you the trouble of reading 90 percent of the responses to the ID position.…This is how losers act just before they lose: arrogant, self-satisfied, too important to be bothered with substantive refutation, and disdainful of their own faults.…The only remaining question is whether Darwinism will exit gracefully, or whether it will go down biting, screaming, censoring, and denouncing to the bitter end."

Tech Central Station contributor **Douglas Kern**, 2005

So the Darwinists are on the defense, and their behavior shows it. According to bestselling author Orson Scott Card, their reactions to criticism are increasingly "illogical, personal, and unscientific." Among other things, they resort to "credentialism" and "expertism." But "real science never has to resort to credentialism. If someone with no credentials at all raises a legitimate question, it is not an answer to point out how uneducated or unqualified the questioner is. In fact, it is pretty much an admission that you don't have an answer, so you want the questioner to go away." And expertism is "the 'trust us, you poor fools' defense. Essentially, the Darwinists tell the general public that we're too dumb to understand.... So who is going to win this argument? Some people bow down before experts; most of us resent the experts who expect us to bow."[13]

Many Darwinists are not merely condescending; they're mean-spirited. Columnist William Rusher writes: "One can't help being a little surprised at the sheer savagery of the evolutionists' attack on intelligent design.... One thinks of scientists as calm, intelligent people, perhaps wearing white smocks, who take on questions to which we don't know the answers, think about them carefully, and test various explanations experimentally until they come up with one that solves the problem.... But that hasn't been the reaction of the evolutionists to intelligent design at all. They have all but bitten themselves in two trying to drive it straight out of the realm of serious discussion."[14]

Tech Central Station contributor Douglas Kern makes fun of the Darwinists' misbehavior: "Ewww...intelligent design people! They're just buck-toothed, Bible-pushing nincompoops with community-college degrees who're trying to sell a gussied-up creationism to a cretinous public! No need to address their

Thoughts are free

"Although they are able to tell us what may be said, they can't tell us what we may think.... Science has never been decided by judicial fiat."

—Cornell IDEA Club founder **Hannah Maxson**, 2005

concerns or respond to their arguments. They are Not Science. They are poopy-heads. There. I just saved you the trouble of reading 90 percent of the responses to the ID position." Kern concludes: "This is how losers act just before they lose: arrogant, self-satisfied, too important to be bothered with substantive refutation, and disdainful of their own faults. . . . The only remaining question is whether Darwinism will exit gracefully, or whether it will go down biting, screaming, censoring, and denouncing to the bitter end."[15]

Obviously, the Darwinists have a growing image problem. But their troubles are just beginning.

Inordinately Well-Funded?

In 2005, Darwinists Richard Dawkins and Jerry Coyne attributed ID's success in part to "loads of tax-free money." Mark Perakh complained about "the inordinately well-funded Center for Science and Culture of the Discovery Institute." The implication is that ID is succeeding only because people are throwing money at it.[16]

The Discovery Institute in Seattle is widely recognized as the international center for intelligent design. It was founded in 1990 as a not-for-profit public policy think tank, and many of its resources go to projects that have nothing to do with ID. The only project that promotes intelligent design is the Center for Science and Culture, and its financial situation is a matter of public record. "Since its founding in 1996," the *New York Times* reported in 2005, "the science center has spent 39 percent of its $9.3 million on research." Much of this went to scholarly books, but "$792,585 financed laboratory or field research in biology, paleontology, or biophysics, while $93,828 helped graduate students." All of the Center's funding came from private donors.[17]

How does this compare with the Darwinists' financial resources? Let's start with the Department of Molecular and Microbiology at George

Mason University (GMU), which drove Dr. Caroline Crocker from biology teaching after she gave a lecture on evidentiary problems with Darwin's theory. According to the department's website, it currently employs four full professors, four associate professors, and seven assistant professors. Statistics for public institutions compiled by the American Association of University Professors suggest that GMU's biology department alone spends over $1.4 million a year on faculty salaries and compensation. That doesn't include substantial amounts spent on support staff, facilities, and research.[18]

GMU is only one of hundreds of public colleges and universities in the United States, all of which are largely tax-supported. This means that the American public is spending hundreds of millions of dollars a year to support Darwinist biology professors. And that's just the tip of the iceberg.

As laid out in Chapter Fifteen, taxpayers are paying for a website that tells students they should believe only religious viewpoints approved by Darwinists. The National Science Foundation, a tax-supported federal agency with an annual budget of more than $5.5 billion, supplied the grant—along with another $2.8 million for pro-Darwin propaganda. Recently, Republican senator Kay Bailey Hutchison of Texas questioned why the NSF is funding social science research in other countries. Maybe it's time for Congress to take a closer look at the agency's entire budget.[19]

The NSF isn't the only offender. The National Cancer Institute, part of the $28 billion a year National Institutes of Health, used taxpayers' hard-earned money—in the name of curing cancer—to pay for a recently published hypothesis about the molecular phylogeny of cats! Inordinately well-funded, indeed.[20]

So Discovery Institute's Center for Science and Culture receives less than $2 million a year in voluntary donations, while Darwinists have their hands in a multi-billion-dollar cookie jar that is refilled every year by contributions that are anything but voluntary. A large majority of the American people reject Darwinism, yet they are forced to support it with their tax money. At the same time, Darwinists treat their benefactors with utter contempt. This is an inherently unstable situation, and it's one more reason why Darwinism will lose.

ID as a Scientific Research Program

A new paradigm succeeds only if it leads to fruitful new research. There are two ways ID can guide scientific research. First, it can suggest theoretical or experimental tests to determine whether certain things are better explained by intelligent design or Darwinian evolution. Second, it can serve as the basis for testable new hypotheses that are unlikely to have emerged from a Darwinian perspective.

In the first category, ID theorist William Dembski has listed various "research themes that may prove helpful to scientists who are trying to find a way to contribute usefully to intelligent design research." Max Planck Institute biologist Wolf-Ekkehard Lönnig sees ID as "the beginning of entirely new research programs" that include tests of irreducible and specified complexity in genetics, anatomy, and physiology.[21]

Adding insult to injury

"Real science never has to resort to credentialism. If someone with no credentials raises a legitimate question, it is not an answer to point out how uneducated or unqualified the questioner is. In fact, it is pretty much an admission that you don't have an answer, so you want the questioner to go away."

—Author **Orson Scott Card**, 2005

On the theoretical level, biochemist Michael Behe and physicist David W. Snoke have tested mathematically the Darwinian hypothesis that "a major route of evolutionary innovation" is "through gene duplication." According to the hypothesis, the original copy of a gene continues to perform its functions while a duplicate copy undergoes random mutations that are unselected until it acquires a new function. Behe and Snoke found that gene duplication can lead to new functions if only one such mutation is needed, but "this conceptually simple pathway for developing new functions is problematic when multiple mutations are required."[22]

On the experimental level, molecular biologist Douglas Axe has tested the Darwinian hypothesis that proteins readily accept random changes to their sequences with no adverse effect. This is necessary if mutation and selection are to have the creative power that Darwinism attributes to them. But when Axe examined the intermediates between two similar enzymes that perform the same job he found that they all lacked biological function. Axe further examined this surprising sensitivity to mutations by randomly replacing clusters of amino acids and testing the resulting proteins for function. He found that the "prevalence of sequences performing a specific function" is very much lower than what would be needed for Darwinian evolution, and he concluded that functional proteins "require highly extraordinary sequences."[23]

In the second category, some scientists are using ID heuristically to develop new hypotheses. For example, Darwinism suggests that pathogenic viruses should long ago have evolved resistance to solar ultraviolet radiation (UV). Yet studies in Brazil by atmospheric scientist Forrest M. Mims III show that most airborne bacteria are quickly suppressed by even small doses of UV, and he regards this as evidence for design. Mims reasons that if ID is true, flu viruses should also be susceptible to UV from sunlight, and he published a prediction that Avian influenza could be "controlled by a substantial reduction" in smoke from regional burning in Southeast Asia that would allow UV from natural sunlight "to suppress the virus before infection occurs."[24]

In 2003, I used intelligent design to develop a hypothesis about centrioles, microscopic structures in animal cells that look like tiny turbines. There are no evolutionary intermediates to support a Darwinian explanation for the origin of centrioles, and Darwinists have been relatively uninterested in them—especially in their resemblance to miniature machines. From an ID perspective, however, centrioles may have been *designed* to function as tiny turbines. If so, their occasional malfunction during cell division could be an early step in the origin of cancer. The hypothesis has been published in a non-Darwinian biology journal.[25]

Like all predictions, these run the risk of being proved wrong. But it's also possible that they—or other ID-based predictions—will be proved right. ID theorists are now hard at work doing scientific experiments to test these and other hypotheses. Unfortunately, most of this research has to be done in secret—not because it is unethical or illegal, but because if Darwinists learn about it they will try to shut it down. In 2005 and 2006, ID scientists held a series of meetings to present their research findings to each other. The meetings were private and unpublicized, because some of the participants stood to lose their jobs if their involvement with ID became public. Nevertheless, ID-guided research is progressing rapidly.

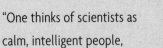

Bring in the men in white coats

"One thinks of scientists as calm, intelligent people, perhaps wearing white smocks, who take on questions to which we don't know the answers, think about them carefully, and test various explanations experimentally until they come up with one that solves the problem.... But that hasn't been the reaction of the evolutionists to intelligent design at all. They have all but bitten themselves in two trying to drive it straight out of the realm of serious discussion."

—Columnist **William Rusher**, 2005

Websites You're Not Supposed to Visit:

http://darwinanddesign.com

http://dcsociety.org

http://pos-darwinista.blogspot.com

Why Intelligent Design Will Win

Darwinists have long claimed that opposition to their view is largely restricted to biblical fundamentalists in the American South. But ID is not biblical fundamentalism, and it is not restricted to the U.S. It is an international phenomenon, and its popularity overseas is growing. In April 2005, *Le Monde* carried an article about its increasing influence among French students. In August 2005, almost seven hundred people from eighteen nations gathered in Prague for a conference on "Darwin and Design." Societies for the study of ID are now active in countries such as Japan and Brazil.[26]

But the most significant sociological fact about intelligent design is its skyrocketing appeal among young people. Intelligent Design and Evolution Awareness (IDEA) clubs, the first of which was founded by students at the University of California–San Diego in 1999, are springing up on college campuses all over the world. There are now clubs in Canada, Kenya, Ukraine, and the Philippines, not to mention several dozen in the U.S. There is a chapter at George Mason University, where Caroline Crocker used to teach. There is also a chapter at Cornell University, where in 2005 Hunter Rawlings warned the trustees and faculty against ID's "assault on science and reason."

Cornell's IDEA club was founded by math and chemistry major Hannah Maxson. She told the *Chicago Tribune:* "In my opinion, both intelligent design and Darwinian evolution are science." After the Dover, Pennsylvania, school board was prohibited from mentioning intelligent design or disparaging Darwinism, Maxson wrote to the *Ithaca Journal*: "Although they are able to tell us what may be said, they can't tell us what we may think. . . . Science has never been decided by judicial fiat."[27]

So a growing number of bright young men and women have the courage to question Darwinism, study intelligent design, and follow the evidence wherever it leads. They know they are in the midst of a major scientific revolution. And the future belongs to them.

~~~~~~~~~~~~~~~~~

ACKNOWLEDGMENTS

I could not have written this book without help from many people. I am indebted most of all to Phillip and Kathie Johnson, without whose inspiration and support this book would never have been written. Many others who helped me must remain anonymous, for reasons that will be clear throughout the book (especially Chapter Sixteen). Of course, the people who helped me may not agree with my conclusions, for which I alone am responsible. Nevertheless, I list here (in alphabetical order) those I can acknowledge publicly: Norris Anderson, Douglas Axe, Michael J. Behe, Francis J. Beckwith, David Berlinski, Nancy Bryson, Larry Caldwell, John Calvert, Bruce Chapman, Seth L. Cooper, Caroline Crocker, Rob Crowther, Paula Currall, Roger DeHart, William A. Dembski, David K. DeWolf, Robert A. DiSilvestro, Michael R. Egnor, Brian Gage, George Gilder, Guillermo Gonzalez, William S. Harris, Fred Heeren, Roland F. Hirsch, Linda Holloway, Cornelius G. Hunter, Tony Jelsma, Casey S. Luskin, Joe Manzari, Timothy and Lydia McGrew, Stephen C. Meyer, Forrest M. Mims III, Scott A. Minnich, Kate Morse, Paul A. Nelson, Lincoln C. Olsen, Nancy Pearcey, Keith Pennock, Del Ratzsch, Jay W. Richards, Nils Roll-Hansen, Marcus R. Ross, Mark Ryland, Jody F. Sjogren, Philip S. Skell, Frederick N. Skiff, Richard M. v. Sternberg, Larry Taylor, Stephen Thompson, Mark A. Toleman, Richard Weikart, Lucy P. Wells, John G. West, and Jonathan Witt.

NOTES

Chapter One:

Wars and Rumors

1. Claudia Wallis, "The Evolution Wars," *Time*, August 15, 2005.

2. Richard Monastersky, "On the Front Lines in the War Over Evolution," *Chronicle of Higher Education*, March 10, 2006.

3. Pat Shipman, "Being Stalked by Intelligent Design," *American Scientist,* November/December 2005. Available online (June 2006) at: http://www.americanscientist.org/template/AssetDetail/assetid/47366?&print=yes. Marshall Berman, "Intelligent Design: The New Creationism Threatens All of Science and Society," *APS News Online* (October 2005). Available online (June 2006) at: http://www.aps.org/apsnews/1005/100518.cfm. Marshall Berman, "Intelligent Design Creationism: A Threat to Society, Not Just Biology," *American Biology Teacher*, 65 (November/December 2003). Holly Foster, "Brown University Professor, Author Ken Miller Lectures on Evolution, Intelligent Design: The Future of Science Is at Stake, Miller Noted," *Hamilton College News,* March 1, 2006. Available online (June 2006) at: http://www.hamilton.edu/news/more_news/display.cfm?ID=10433.

4. Eugenie C. Scott, "Dealing with Anti-Evolutionism," University of California–Berkeley Museum of Paleontology website. Available online (June 2006) at: http://www.ucmp.berkeley.edu/fosrec/Scott1.html.

5. Keith Stewart Thomson, "The Meanings of Evolution," *American Scientist* 70 (September/October 1982). Peter J. Bowler, *Evolution: The History of an Idea,* Revised Edition (Berkeley: University of California Press, 1989).

6. Charles Darwin, *The Origin of Species*, Introduction and Conclusion.

7. U.S. National Academy of Sciences, *Teaching About Evolution and the Nature of Science* (Washington, DC: National Academies Press, 1998), Chapters 1 and 5. Available online (June 2006) at: http://www.nap.edu/readingroom/books/evolution98.

8. Francis Darwin (editor), *Life and Letters of Charles Darwin* (New York: Appleton, 1887), Volume II, 202. Marc W. Kirschner and John C. Gerhart, *The Plausibility of Life* (New Haven, CT: Yale University Press, 2005), 46–50.

9. Asa Gray, *Darwiniana* (New York: Appleton, 1876), 148–57.

10. Francis Darwin (editor), *Life and Letters of Charles Darwin*, Volume II, 80. Charles Darwin, *The Variation of Animals and Plants Under Domestication* (London: Orange Judd, 1868), Volume II, 514–16.

11. Francis Darwin (editor), *Life and Letters of Charles Darwin*, Volume I, 278–85; Volume II, 105–06. Francis Darwin and A. C. Seward (editors), *More Letters of Charles Darwin* (New York: Appleton, 1903), Volume I, 126, 321. Jonathan Wells, "Charles Darwin on the Teleology of Evolution," *International Journal on the Unity of the Sciences* 4 (Summer 1991).

12. Horace Freeland Judson, *The Eighth Day of Creation: The Makers of the Revolution in Biology* (New York: Simon and Schuster, 1979), 216–17.

13. Richard Dawkins, *The Blind Watchmaker: Why the Evidence of Evolution Reveals a Universe without Design* (New York: W. W. Norton, 1986), 1.

14. Francis Darwin and Seward (editors), *More Letters of Charles Darwin*, Volume I, 321. Charles Darwin, *The Origin of Species*, Conclusion.

15. Charles Coulston Gillispie, *Genesis and Geology* (New York: Harper & Row, 1959). Peter J. Bowler, *Evolution: The History of an Idea,* 218. Jonathan Wells, *Charles Hodge's Critique of Darwinism: An Historical-Critical Analysis of Concepts Basic to the 19th Century Debate* (Lewiston, NY: Edwin Mellen Press, 1988).

16. Charles B. Thaxton, Walter L. Bradley, and Roger L. Olsen, *The Mystery of Life's Origin* (Dallas, TX: Lewis and Stanley, 1984). Michael Denton, *Evolution: A Theory in Crisis* (Bethesda, MD: Adler & Adler, 1985). Phillip E. Johnson, *Darwin On Trial*, Second Edition (Downer's Grove, IL: InterVarsity Press, 1993). The meeting (at Pajaro Dunes) near Monterey is described in the video *Unlocking the Mystery of Life* (La Habra, CA: Illustra Media, 2002). For a brief history of ID, see Jonathan Witt, "The Origin of Intelligent Design: A Brief History of the Scientific Theory of Intelligent Design," Dis-

covery Institute. Available online (June 2006) at: http://www.discovery.org/scripts/viewDB/index.php?command=view&id=3207.

17. Stephen C. Meyer, "Intelligent Design Is Not Creationism," *Daily Telegraph,* January 28, 2006. Available online (June 2006) at: http://www.telegraph.co.uk/opinion/main.jhtml?xml=/opinion/2006/01/28/do2803.xml&sSheet=/opinion/2006/01/28/ixop.html. Henry Morris, "Intelligent Design and/or Scientific Creationism," *Acts & Facts* 35 (April 2006): Back to Genesis #208. Available online (June 2006) at: http://www.icr.org/index.php?module=articles&action=view&ID=2708. Carl Wieland, "AiG's views on the Intelligent Design Movement," *Answers in Genesis*, August 30, 2002. Available online (June 2006) at: http://www.answersingenesis.org/docs2002/0830_IDM.asp. *Sunday Times Review*, "In the Beginning There Was Something," *Sunday Times*, December 19, 2004. Available online (June 2006) at: http://www.timesonline.co.uk/article/0,,2092-1408276_1,00.html.

18. "Questions about Intelligent Design," Discovery Institute, 2006. Available online (June 2006) at: http://www.discovery.org/csc/topQuestions.php.

19. Eugenie C. Scott, "Dealing with Anti-Evolutionism," University of California–Berkeley Museum of Paleontology website. Available online (June 2006) at: http://www.ucmp.berkeley.edu/fosrec/Scott2.html.

20. Ibid.

21. Jerry Adler, "Evolution of a Scientist," *Newsweek*, November 28, 2005.

22. J. A. Simpson and E. S. C. Weiner, *Oxford English Dictionary*, Second Edition, Vol. 4, (Oxford: Clarendon Press, 1989), 257. Asa Gray, *Darwiniana: Essays and Reviews Pertaining to Darwinism* (New York: D. Appleton, 1876). Alfred Russel Wallace, *Darwinism: An Exposition of the Theory of Natural Selection, With Some of Its Applications* (London: Macmillan, 1889). Ernst Mayr, *The Growth of Biological Thought* (Cambridge, MA: Harvard University Press, 1982), 116–17, 505. Stephen Jay Gould, *The Structure of Evolutionary Theory* (Cambridge, MA: Harvard University Press, 2002). A. M. Silverstein, "Darwinism and Immunology," *Nature Immunology* 4 (2003), 3–6. G. S. Levit, U. Hossfeld, and L. Olsson, "The Integration of Darwinism and Evolutionary Morphology," *Journal of Experimental Zoology B: Molecular and Developmental Evolution* 302 (2004): 343–54.

23. Robert T. Pennock (editor), *Intelligent Design Creationism and Its Critics* (Cambridge, MA: MIT Press, 2001). Matt Ridley, "Intelligent Design Lacks

Intelligence," *Daily Telegraph*, January 30, 2006. Available online (June 2006) at: http://www.telegraph.co.uk/opinion/main.jhtml?xml=/opinion/2006/01/30/dt3001.xm. Richard Ostling, "Ohio School Board Debates Teaching 'Intelligent Design'," *Washington Post*, March 14, 2002. Available online (June 2006) at: http://www.discovery.org/scripts/viewDB/index.php?command=view&id=1140.

Chapter Two:
What the Fossil Record *Really* Says

1. National Academy of Sciences, *Teaching About Evolution and the Nature of Science* (Washington, DC: National Academy Press, 1998), Chapter 2. Available online (June 2006) at: http://www.nap.edu/readingroom/books/evolution98/. National Academy of Sciences, *Science and Creationism*, Second Edition (Washington, DC: National Academy Press, 1999), Evidence Supporting Biological Evolution. Available online (June 2006) at: http://fermat.nap.edu/html/creationism/.

2. Kenneth R. Miller and Joseph S. Levine, *Prentice Hall Biology* (Upper Saddle River, NJ: Pearson Education, 2002), 382.

3. Charles Darwin, *The Origin of Species*, Chapters IV, X.

4. James W. Valentine, Stanley M. Awramik, Philip W. Signor, and Peter M. Sadler, "The Biological Explosion at the Precambrian-Cambrian Boundary," *Evolutionary Biology* 25 (1991): 279–356. Jeffrey S. Levinton, "The Big Bang of Animal Evolution," *Scientific American* 267, (November 1992), 84–91. J. Madeleine Nash, "When Life Exploded," *Time*, December 4, 1995.

5. Darwin, *The Origin of Species*, Chapter X.

6. J. William Schopf & Bonnie M. Packer, "Early Archean (3.3 Billion to 3.5 Billion Year-Old) Microfossils from Warrawoona Group, Australia," *Science* 237 (1987): 70–73. Simon Conway Morris, *The Crucible of Creation* (Oxford: Oxford University Press, 1998), 2. J. William Schopf, "The Early Evolution of Life: Solution to Darwin's Dilemma," *Trends in Ecology and Evolution* 9 (1994): 375–77.

7. Valentine, et al., "The Biological Explosion at the Precambrian-Cambrian Boundary," 281, 318.

8. Ibid., 294. Harry B. Whittington, *The Burgess Shale* (New Haven, CT: Yale University Press, 1985), 131.

9. James W. Valentine, *On the Origin of Phyla* (Chicago: The University of Chicago Press, 2004), xxiii.

10. Michael J. Benton, *Vertebrate Palaeontology*, Second Edition (London: Chapman & Hall, 1997). J. G. M. Thewissen and E. M. Williams, "The Early Radiations of Cetacea (Mammalia): Evolutionary Pattern and Developmental Correlations," *Annual Review of Ecology and Systematics* 33 (2002): 73–90.

11. Percival Davis, Dean H. Kenyon, and Charles B. Thaxton, *Of Pandas and People: The Central Question of Biological Origins,* Second Edition (Dallas, TX: Haughton Publishing, 1993), 96, 101–02.

12. J. G. M. Thewissen, S. T. Hussain, and M. Arif, "Fossil Evidence for the Origin of Aquatic Locomotion in Archaeocete Whales," *Science* 263 (1994): 210–12. Annalisa Berta, "What Is a Whale?" *Science* 263 (1994), 180–81. Philip D. Gingerich, S. Mahmood Raza, Muhammad Arif, Mohammad Anwar, and Xiaoyuan Zhou, "New Hale from the Eocene of Pakistan and the Origin of Cetacean Swimming," *Nature* 368 (1994): 844–47. Ashby Camp, "The Overselling of Whale Evolution," *The True.Origins Archive,* 1998. Available online (June 2006) at: http://www.trueorigin.org/whales.asp.

13. Stephen Jay Gould, "Hooking Leviathan By Its Past," *Natural History* 103 May (1994), 8–14.

14. Kevin Padian, "The Tale of the Whale," National Center for Science Education Resources. Available online (June 2006) at: http://www.ncseweb.org/resources/rncse_content/vol17/2010_the_tale_of_the_whale_1 2_30_1899.asp.

15. Robert L. Carroll, *Patterns and Processes of Vertebrate Evolution* (Cambridge: Cambridge University Press, 1997), 331. J. G. M. Thewissen, "A Family Tree of Whales," *Whale Origins Research,* 2006. Available online (June 2006) at: http://darla.neoucom.edu/DEPTS/ANAT/Thewissen/whale_origins/index.html.

16. Tim Berra, *Evolution and the Myth of Creationism* (Stanford, CA: Stanford University Press, 1990), 117–19. Phillip E. Johnson, *Defeating Darwinism by Opening Minds* (Downers Grove, IL: InterVarsity Press, 1997), 62–63.

17. Gareth Nelson, "Presentation to the American Museum of Natural History," 1969. In David M. Williams and Malte C. Ebach, "The Reform of

Palaeontology and the Rise of Biogeography—25 years after 'ontogeny, phylogeny, palaeontology and the biogenetic law' (Nelson, 1978)," *Journal of Biogeography* 31 (2004), 685–712, 709.

18. Henry Gee, *In Search of Deep Time: Beyond the Fossil Record to a New History of Life* (New York: The Free Press, 1999), 32, 113–17.

19. *Teaching Science in a Climate of Controversy* (Ipswich, MA: American Scientific Affiliation, 1986), 56–63.

20. Kevin Padian and Kenneth D. Angielczyk, "Are There Transitional Forms in the Fossil Record?" 47–82 in P. H. Kelley, J. R. Bryan, and T. A. Hansen (editors), *The Evolution-Creationism Controversy II: Perspectives on Science, Religion and Geological Education* (Fayetteville, AR: Paleontological Society, 1999).

21. Kathy Canavan, "Comedian Lewis Black Decries the 'Politically Correct'," *University of Delaware Daily*, November 15, 2002. Available online (June 2006) at: http://www.udel.edu/PR/UDaily/01-02/lblack111502.html.

Chapter Three:
Why You Didn't "Evolve" in Your Mother's Womb

1. Charles Darwin, *The Origin of Species*, Chapter XIV. Charles Darwin, *The Descent of Man*, Chapter I. September 10, 1860, letter to Asa Gray, in Francis Darwin (editor), *The Life and Letters of Charles Darwin* (New York: D. Appleton & Company, 1896), Vol. II, 131.

2. Arthur Henfrey & Thomas H. Huxley (editors), *Scientific Memoirs: Selected from the Transactions of Foreign Academies of Science and from Foreign Journals: Natural History* (London, 1853; reprinted 1966 by Johnson Reprint Corporation, New York), 214.

3. Timothy Lenoir, *The Strategy of Life* (Chicago: The University of Chicago Press, 1982), 258.

4. Frederick B. Churchill, "The Rise of Classical Descriptive Embryology," 1–29 in Scott F. Gilbert (editor), *A Conceptual History of Modern Embryology* (Baltimore, MD: The Johns Hopkins University Press, 1991), 19–20.

5. Jonathan Wells, "Haeckel's Embryos and Evolution: Setting the Record Straight," *American Biology Teacher* 61 (May 1999): 345–49. Jonathan Wells, *Icons of Evolution: Why Much of What We Teach About Evolution Is Wrong* (Washington, DC: Regnery Publishing, 2002), Chapter 5.

6. J. Assmuth and Ernest R. Hull, *Haeckel's Frauds and Forgeries* (Bombay: Examiner Press, 1915). M. K. Richardson, J. Hanken, M. L. Gooneratne, C. Pieau, A. Raynaud, L. Selwood, and G. M. Wright, "There is no highly conserved embryonic stage in the vertebrates: implications for current theories of evolution and development," *Anatomy & Embryology* 196 (1997): 91–106. Richardson was quoted in Elizabeth Pennisi, "Haeckel's Embryos: Fraud Rediscovered," *Science* 277 (1997): 1435.

7. Stephen Jay Gould, "Abscheulich! Atrocious!" *Natural History* (March, 2000), 42–49, 44–46. Cecie Starr and Ralph Taggart, *Biology: The Unity and Diversity of Life,* Tenth Edition (Belmont, CA: Thomson Learning, 2004), 315. Joseph Raver, Biology: *Patterns and Processes of Life* (Dallas, TX: J. M. LeBel Publishers, 2004), 100. Donald Voet and Judith G. Voet, *Biochemistry,* Third Edition (Hoboken, NJ: John Wiley & Sons, 2004), 4.

8. Jerry A. Coyne, "Creationism by stealth," *Nature* 410 (2001): 745. Eugenie Scott, interview in "Icons of Evolution," Coldwater Media, 2002. Douglas J. Futuyma, *Evolution* (Sunderland, MA: Sinauer Associates, 2005), 535.

9. Ibid.

10. Lewis Wolpert, *The Triumph of the Embryo* (Oxford: Oxford University Press, 1991), 12.

11. Adam Sedgwick, "On the Law of Development Commonly Known as von Baer's Law; and on the Significance of Ancestral Rudiments in Embryonic Development," *Quarterly Journal of Microscopical Science* 36 (1894), 35–52.

12. William W. Ballard, "Problems of gastrulation: real and verbal," *BioScience* 26 (1976): 36–39. Richard P. Elinson, "Change in Developmental Patterns: Embryos of Amphibians with Large Eggs," 1–21 in R. A. Raff & E. C. Raff (editors), *Development as an Evolutionary Process*, Vol. 8 (New York: Alan R. Liss, 1987).

13. Rudolf A. Raff, *The Shape of Life: Genes, Development, and the Evolution of Animal Form* (Chicago: The University of Chicago Press, 1996), 195, 208–09.

14. Kenneth R. Miller and Joseph S. Levine, *Prentice Hall Biology* (Upper Saddle River, NJ: Pearson Education, 2002), 385.

15. Gregory Wray, "Punctuated Evolution of Embryos," *Science* 267 (1995), 1115–16. Rudolf A. Raff, *The Shape of Life: Genes, Development,* *and the Evolution of Animal Form* (Chicago: The University of Chicago Press, 1996), 211.

16. Brian K. Hall, *Evolutionary Developmental Biology* (London: Chapman & Hall, 1992), Preface. Peter W. H. Holland, "The Future of Evolutionary Developmental Biology," *Nature* 402 Suppl. (1999), C41–C44. Corey S. Goodman and Bridget C. Coughlin, "The Evolution of Evo-Devo Biology," *Proceedings of the National Academy of Sciences USA* 97 (2000), 4424–25.

17. Georg Halder, Patrick Callaerts, and Walter J. Gehring, "Induction of Ectopic Eyes by Targeted Expression of the *Eyeless* Gene in Drosophila," *Science* 267, 1995, 1788–92.

18. Giuseppe Sermonti, *Why Is a Fly Not a Horse?* (Seattle, WA: Discovery Institute Press, 2005).

19. Sean B. Carroll, "Endless Forms Most Beautiful: A New Revolution in Biology," *Skeptical Inquirer* 29:6 (November/December 2005), 48–53.

20. E. B. Lewis, "A Gene Complex Controlling Segmentation in *Drosophila*," *Nature* 276 (1978), 565–70. E. B. Lewis, "Regulation of the Genes of the Bithorax Complex in *Drosophila*," *Cold Spring Harbor Symposia on Quantitative Biology* 50, (1985), 155–64.

21. J. Fernandes, S. E. Celniker, E. B. Lewis & K. VijayRaghavan, "Muscle Development in the Four-Winged *Drosophila* and the Role of the *Ultrabithorax* Gene," *Current Biology* 4, (1994), 957–64. See also Jonathan Wells, *Icons of Evolution*, Chapter 9.

22. "First Genetic Evidence Uncovered of How Major Changes in Body Shapes Occurred During Early Animal Evolution," UCSD News Release, February 6, 2002. Available online (January 2006) at: http://ucsdnews.ucsd.edu/newsrel/science/mchox.htm.

23. Matthew Ronshaugen, Nadine McGinnis, and William McGinnis, "Hox Protein Mutation and Macroevolution of the Insect Body Plan," *Nature* 415, 2002, 914–17. For an analysis, see Jonathan Wells, "Mutant Shrimp?—A Correction," February 11, 2002. Available online (January 2006) at: http://www.discovery.org/scripts/viewDB/index.php?command=view&id=1118.

24. Marc W. Kirschner and John C. Gerhart, *The Plausibility of Life: Resolving Darwin's Dilemma* (New Haven, CT: Yale University Press, 2005), 237.

25. Arhat Abzhanov, Meredith Protas, B. Rosemary Grant, Peter R. Grant, and Clifford J. Tabin, "*Bmp4* and Morphological Variation of Beaks in Darwin's Finches," *Science* 305, 2004, 1462–65. H. Lisle Gibbs and Peter R. Grant, "Oscillating Selection on Darwin's Finches," *Nature* 327, 1987, 511–13. Jonathan Weiner, *The Beak of the Finch* (New York: Vintage Books, 1994), 104–05, 176.

Chapter Four:

What Do Molecules Tell Us about Our Ancestors?

1. U.S. National Academy of Sciences, *Science and Creationism*, Second Edition (Washington, DC: National Academy Press, 1999), Section on "Evidence Supporting Biological Evolution." Available online (June 2006) at: http://newton.nap.edu/html/creationism/evidence.html.

2. Douglas J. Futuyma, *Evolution* (Sunderland, MA: Sinauer Associates, 2005), 528.

3. Emile Zuckerkandl and Linus Pauling, "Molecules as Documents of Evolutionary History," *Journal of Theoretical Biology* 8, 1965, 357–66. Hendrik N. Poinar et al., "Metagenomics to Paleogenomics: Large-Scale Sequencing of Mammoth DNA," *Science* 311, 2000, 392–94. Stanford University, "Ancient DNA Provides Clues to the Evolution of Social Behavior," *Science Daily*, April 22, 2006. Available online (June 2006) at: http://www.sciencedaily.com/releases/2006/04/060421110553.htm.

4. Leigh Van Valen, "Deltatheridia, a New Order of Mammals," *Bulletin of the American Museum of Natural History* 132, 1966, 1–126. Leigh Van Valen, "Monophyly or Diphyly in the Origin of Whales," *Evolution* 22 1968, 37–41.

5. Dennis Normile, "New Views of the Origins of Mammals," *Science* 281, 1998, 774–75. Richard Monastersky, "The Whale's Tale: Research on Whale Evolution," *Science News*, November 6, 1999. Available online (June 2006) at: http://www.findarticles.com/p/articles/mi_m1200/is_19_156/ai_57828404.

6. John E. Heyning, "Whale Origins—Conquering the Seas," *Science* 283 (1999): 943, 1642–43. Maureen A. O'Leary and Jonathan H. Geisler, "The Position of Cetacea within Mammalia: Phylogenetic Analysis of Morphological Data from Extinct and Extant Taxa," *Systematic Biology* 48 (1999) 455–90. Abstract available online (June 2006) at:

http://www.ncbi.nlm.nih.gov/entrez/query.fcgi?CMD=search&DB= pubmed. Zhexi Luo, "In Search of the Whales' Sisters," *Nature* 404 (2000) 235–39. John Gatesy and Maureen A. O'Leary, "Decipering Whale Origins with Molecules and Fossils," *Trends in Ecology and Evolution* 16 (2001), 562–70.

7. Philip D. Gingerich, Munir ul Haq, Iyad S. Zalmout, Intizar Hussain Khan, and M. Sadiq Malkani, "Origin of Whales from Early Artiodactyls: Hands and Feet of Eocene Protocetidae from Pakistan," *Science* 293 (2001), 2239–42. J. G. M. Thewissen, E. M. Williams, L. J. Roe, and S. T. Hussain, "Skeletons of terrestrial ceataceans and the relationship of whales to artiodactyls," *Nature* 413 (2001): 277–81. Christian de Muizon, "Walking with whales," *Nature* 413 (2001): 259–60. Kenneth D. Rose, "The Ancestry of Whales," *Science* 293 (2001): 2216–17.

8. Anna Marie A. Aguinaldo, James M. Turbeville, Lawrence S. Linford, Maria C. Rivera, James R. Garey, Rudolf A. Raff, and James A. Lake, "Evidence for a clade of nematodes, arthropods and other molting animals," *Nature* 387 (1997): 489–93.

9. Michael Lynch, "The Age and Relationships of the Major Animal Phyla," *Evolution* 53 (1999): 319–25.

10. André Adoutte, Guillaume Balavoine, Nicolas Lartillot, Olivier Lespinet, Bejamin Purd'homme, and Renaud de Rosa, "The new animal phylogeny: Reliability and implications," *Proceedings of the National Academy of Sciences USA* 97 (2000), 4453–56. Available online (June 2006) at: http://www.pnas.org/cgi/reprint /97/9/4453. Jaime E. Blair, Kazuho Ikeo, Takashi Gojobori, and S. Blair Hedges, "The evolutionary position of nematodes," *Biomed Central Evolutionary Biology* 2 (2002), 7. Available online (June 2006) at: http://www.biomedcentral.com/content/pdf/1471-2148-2-7.pdf.

11. Yuri I. Wolf, Igor B. Rogozin, and Eugene V. Koonin, "Coelomata and Not Ecdysozoa: Evidence From Genome-Wide Phylogenetic Analysis," *Genome Research* 14 (2004): 29–36. Available online (June 2006) at: http://www.genome.org/cgi/reprint/14/1/29.pdf. Hervé Philippe, Nicolas Lartillot, and Henner Brinkmann, "Multigene Analysis of Bilaterian Animals Corroborate the Monophyly of Ecdysozoa, Lophotrochozoa, and Protostomia," *Molecular Biology and Evolution* 22 (2005): 1246–53.

12. Martin Jones and Mark Blaxter, "Animal roots and shoots," *Nature* 434 (2005): 1076–77. Antonis Rokas, Dirk Krüger, and Sean B. Carroll, "Animal Evolution and the Molecular Signature of Radiations Compressed in Time," *Science* 310 (2005): 1933–38.

13. James A. Lake, Ravi Jain, and Maria C. Rivera, "Mix and Match in the Tree of Life," *Science* 283 (1999): 2027–28. Hervé Philippe and Patrick Forterre, "The Rooting of the Universal Tree of Life Is Not Reliable," *Journal of Molecular Evolution* 49 (1999): 509–23.

14. Carl Woese, "The universal ancestor," *Proceedings of the National Academy of Sciences USA* 95 (1998): 6854–59.

15. W. Ford Doolittle, "Phylogenetic Classification and the Universal Tree," *Science* 284 (1999): 2124–28. W. Ford Doolittle, "Lateral Genomics," *Trends in Biochemical Sciences* 24 (1999): M5–M8. W. Ford Doolittle, "Uprooting the Tree of Life," *Scientific American* 282 (February, 2000): 90–95.

16. Patrick Forterre and Hervé Philippe, "Where is the root of the universal tree of life," *BioEssays* 21 (1999): 871–79. Patrick Forterre and Hervé Philippe, "The Last Universal Common Ancestor (LUCA), Simple or Complex?" *Biological Bulletin* 196 (1999), 373–77.

17. Carl Woese, "On the evolution of cells," *Proceedings of the National Academy of Sciences USA* 99 (2002): 8742–47. Carl R. Woese, "A New Biology for a New Century," *Microbiology and Molecular Biology Reviews* 68 (2004): 173–86. Eric Bapteste, Yan Boucher, Jessica Leigh, and W. Ford Doolittle, "Phylogenetic Reconstruction and Lateral Gene Transfer," *Trends in Microbiology* 12 (2004), 406–11. E. Bapteste, E. Susko, J. Leigh, D. MacLeod, R. L. Charlebois, and W. F. Doolittle, "Do Orthologous Gene Phylogenies Really Support Tree-Thinking?" *Biomed Central Evolutionary Biology* 5 (2005), 33. Available online (June 2006) at: http://www.biomedcentral.com/content/pdf/1471-2148-5-33.pdf. S. L. Baldauf, "The Deep Roots of Eukaryotes," *Science* 300 (2003), 1703–06.

18. Maria C. Rivera and James A. Lake, "The Ring of Life Provides Evidence for a Genome Fusion Origin of Eukaryotes," *Nature* 431 (2004), 152–55. William Martin and T. Martin Embley, "Early Evolution Comes Full Circle," *Nature* 431 (2004), 134–37.

19. Dan Graur and William Martin, "Reading the Entrails of Chickens: Molecular Timescales of Evolution and the Illusion of Precision," *Trends in Genetics* 20 (2004), 80–86.

Chapter Five:
The Ultimate Missing Link

1. Ernst Mayr, *The Growth of Biological Thought* (Cambridge, MA: Harvard University Press, 1982), 403.

2. Charles Darwin, *The Origin of Species*, Introduction. Keith Stewart Thomson, "Natural Selection and Evolution's Smoking Gun," *American Scientist* 85 (1997), 516–18.

3. Catherine A. Callaghan, "Instances of Observed Speciation," *American Biology Teacher* 49 (1987), 34–36. Joseph Boxhorn, "Observed Instances of Speciation," *The Talk.Origins Archive*, September 1, 1995. Available online (January 2006) at: http://www.talkorigins.org/faqs/faq-speciation.html. Chris Stassen, James Meritt, Annelise Lilje, and L. Drew Davis, "Some More Observed Speciation Events," *The Talk.Origins Archive*, 1997. Available online (January 2006) at: http://www.talkorigins.org/faqs/speciation.html.

4. Arne Müntzing, "Cytogenetic Investigations on Synthetic *Galeopsis tetrahit*," *Hereditas* 16 (1932), 105–54. Justin Ramsey and Douglas W. Schemske, "Neopolyploidy in Flowering Plants," *Annual Review of Ecology and Systematics* 33 (2002), 589–639.

5. Douglas J. Futuyma, *Evolution* (Sunderland, MA: Sinauer Associates, 2005), 398.

6. Jerry A. Coyne and H. Allen Orr, *Speciation* (Sunderland, MA: Sinauer Associates, 2004), 25–39.

7. J. M. Thoday and J. B. Gibson, "Isolation by Disruptive Selection," *Nature* 193 (1962), 1164–66. J. M. Thoday and J. B. Gibson, "The Probability of Isolation by Disruptive Selection," *American Naturalist* 104 (1970), 219–30. Coyne and Orr, *Speciation,* 138.

8. Theodosius Dobzhansky and Olga Pavlovsky, "Spontaneous Origin of an Incipient Species in the *Drosophila Paulistorum* Complex," *Proceedings of the National Academy of Sciences* 55 (1966), 727–33. Coyne and Orr, *Speciation,* 138.

9. James R. Weinberg, Victoria R. Starczak, and Daniele Jörg, "Evidence for Rapid Speciation Following a Founder Event in the Laboratory," *Evolu-

tion 46 (1992), 1214–20. Francisco Rodriquez-Trelles, James R. Weinberg, and Francisco J. Ayala, "Presumptive Rapid Speciation After a Founder Event in a Laboratory Population of *Nereis*: Allozyme Electrophoretic Evidence Does Not Support the Hypothesis," *Evolution* 50 (1996): 457–61.

10. E. Paterniani, "Selection for Reproductive Isolation between Two Populations of Maize, *Zea mays* L.," *Evolution* 23 (1969), 534–47.

11. William R. Rice and George W. Salt, "Speciation via Disruptive Selection on Habitat Preference: Experimental Evidence," *American Naturalist* 131 (1988), 911–17. Coyne and Orr, *Speciation*,138–41.

12. Lynn Margulis and Dorion Sagan, *Acquiring Genomes: A Theory of the Origins of Species* (New York: Basic Books, 2002), 32. Richard Milton, "Observed instances of speciation," *Alternative Science*, 2004. Available online (January 2006) at: http://www.alternativescience.com/talk.origins-speciations.htm. As this book was going to press, *Nature* published a report of speciation by hybridization (without polyploidy) in Central American butterflies. Jesus Mavarez et al., "Speciation by hybridization in Heliconius butterflies," *Nature* 441 (2006), 868–71. If corroborated, this would be the first case of observed speciation in animals. But speciation by hybridization (with or without polyploidy) is not "evolution's smoking gun." Darwin's theory requires that one species split into two that continue to diverge, not that two species combine to make a third with intermediate characteristics.

13. Futuyma, *Evolution*, 401.

14. Gary Hurd, "To the Committee," Review of Proposed Changes to Kansas State Science Standards. Available online (April 2006) at: http://www.ksde.org/outcomes/sciencereviewhurd.pdf.

15. Theodosius Dobzhansky, *Genetics and the Origin of Species*, Reprinted 1982 (New York: Columbia University Press, 1937), 12.

16. Richard Goldschmidt, *The Material Basis of Evolution* (New Haven: Yale University Press, 1940), 8, 396.

17. Scott F. Gilbert, John M. Opitz, and Rudolf A. Raff, "Resynthesizing Evolutionary and Developmental Biology," *Developmental Biology* 173 (1996): 357–72. Sean B. Carroll, "The big picture," *Nature* 409 (2001), 669.

18. David Whitehouse, "Scientists see new species born," *BBC News*, June 9, 2004. Available online (January 2006) at: http://news.bbc.co.uk/2/hi/science/nature/3790531.stm. See also the June 2004 press release from the

University of Arizona, available online (January 2006) at: http://www.newswise.com/p/articles/view/505399/. Laura K. Reed and Therese A. Markow, "Early events in speciation: Polymorphism for Hybrid Male Sterility in *Drosophila*," *Proceedings of the National Academy of Sciences USA* 101 (June 15, 2004), 9009–12. Available online (January 2006) at: http://www.pnas.org/cgi/reprint/101/24/9009.

19. Kenneth R. Miller, "Statement to the Kansas State Board of Education," 2005. Available online (January 2006) at: http://www.ksde.org/outcomes/sciencereviewmiller.pdf. Linn et al. 2004, "Postzygotic isolating factor in sympatric speciation in *Rhagoletis* Flies: Reduced Response of Hybrids to Parental Host-Fruit Odors," *Proceedings of the National Academy of Sciences USA* 101 (2004), 17753–58. Available online (January 2006) at: http://www.pnas.org/cgi/reprint/101/51/17753.

20. Coyne & Orr, *Speciation*, 90–91.

21. Alan Linton, "Scant Search for the Maker," *Times Higher Education Supplement,* April 20, 2001, Book Section, 29. Available online with registration (January 2006) at: http://www.thes.co.uk/search/story.aspx?story_id=72809.

Chapter Six:
Not Even a Theory

1. Stephen Jay Gould, "Evolution as Fact and Theory," *Hen's Teeth and Horse's Toes* (New York: W. W. Norton & Company, 1994), 253–62. Originally published in *Discover*, May 1981. Available online (June 2006) at: http://www.stephenjaygould.org/library/gould_fact-and-theory.html.

2. Douglas J. Futuyma, *Science on Trial* (New York: Pantheon Books, 1983), 166–71.

3. Norris Anderson, "The Alabama Insert: A Call for Impartial Science," *Access Research Network*, May 15, 1996. Available online (June 2006) at: http://www.arn.org/docs/anderson/insert.htm.

4. Stephen Jay Gould, "Introduction," in Carl Zimmer, *Evolution: The Triumph of an Idea* (New York: HarperCollins, 2001), ix–xiv.

5. "Text of Cobb County Disclaimer," Approved by Cobb County Board of Education, Thursday, March 28, 2002. Available online (June 2006) at: http://www.discovery.org/scripts/viewDB/index.php?command=view&id=3362. Art Toalston, "'Balanced education' in science: Atlanta-area school

board holds firm," Baptist Press, September 27, 2002. Available online (June 2006) at: http://www.bpnews.net/bpnews.asp?ID=14337.

6. "Judge: Evolution Stickers Unconstitutional," CNN (January 13, 2005). Available online (June 2006) at: http://www.cnn.com/2005/LAW/01/13/evolution.textbooks.ruling/index.html. "Background Information on Cobb County School District v. Selman," Discovery Institute, December 16, 2005. Available online (June 2006) at: http://www.discovery.org/scripts/viewDB/index.php?command=view&id=2290#Releases.

7. Douglas J. Futuyma, *Evolution* (Sunderland, MA: Sinauer Associates, 2005), xv, 527.

8. Richard Dawkins, "The 'Alabama Insert': A Study in Ignorance and Dishonesty," *Journal of the Alabama Academy of Science,* Vol. 68, No. 1, January, 1997. Available online (June 2006) at: http://www.simonyi.ox.ac.uk/dawkins/WorldOfDawkins-archive/Dawkins/Work/Articles/alabama/1996-04-01alabama.shtml.

9. The Mutations, "Overwhelming Evidence," *Uncommon Descent*, February 8, 2006. Available online (June 2006) at: http://www.uncommondescent.com/index.php/archives/797.

10. Inside Science News Service, "Physics Nobelist Takes Stand on Evolution," American Institute of Physics (2003). Available online (June 2006) at: http://www.aip.org/isns/reports/2003/081.html.

11. American Association of University Professors, "Teaching Evolution," June 15, 2005. Available online (June 2006) at: http://www.aaup.org/statements/Resolutions/TeachingEvolution.htm. AAAS News, "Science, Teachers and Clergy Strengthen Bonds at AAAS Evolution Event," February 20, 2006. Available online (June 2006) at: http://www.aaas.org/news/releases/2006/0220evo.shtml.

12. W. C. Dampier, *A History of Science*, Fourth Edition Reprinted (Cambridge: Cambridge University Press, 1977), 109–13, 182–84. Peter J. Bowler, *Evolution: The History of an Idea*, Revised Edition (Berkeley: University of California Press, 1989), 246.

13. Deborah Jordan Brooks, "Substantial Numbers of Americans Continue to Doubt Evolution as Explanation for Origin of Humans," Gallup News Service, March 5, 2001. Available online (June 2006) at: http://www.unl.edu/rhames/courses/current/creation/evol-poll.htm. "100

Scientists, National Poll Challenge Darwinism," Zogby International, September 25, 2001. Available online (June 2006) at: http://www.zogby.com/ soundbites/ReadClips.dbm?ID=3944. "Nearly Two-thirds of U.S. Adults Believe Human Beings Were Created by God," Harris Poll, July 6, 2005. Available online (June 2006) at: http://www.harrisinteractive.com/ harris_poll/index.asp?PID=581. Michael Foust, "Gallup poll latest to show Americans reject secular evolution," BP News, October 19, 2005. Available online (June 2006) at: http://www.bpnews.net/bpnews.asp?ID=21891. "A Scientific Dissent from Darwinism," Discovery Institute. Available online (June 2006) at: http://www.dissentfromdarwin.org/. NCSE, "Project Steve," National Center for Science Education, February 16, 2003. Available online (June 2006) at: http://www.ncseweb.org/resources/articles/3541_ project_steve_2_16_2003.asp.

14. Charles Darwin, *The Origin of Species*, Chapters 6 and 14.

15. Douglas J. Futuyma, *Evolution* (Sunderland, MA: Sinauer Associates, 2005), 49.

16. Neal C. Gillespie, *Charles Darwin and the Problem of Creation* (Chicago: The University of Chicago Press, 1979), 124–25. Paul A. Nelson, "The role of theology in current evolutionary reasoning," *Biology and Philosophy* 11, October 1996: 493–517.

17. Cornelius G. Hunter, *Darwin's God: Evolution and the Problem of Evil* (Grand Rapids, MI: Brazos Press, 2001), 48–49, 84, 146, 158.

18. Gillespie, *Charles Darwin and the Problem of Creation,* 8, 54, 115, 146–47.

19. Phillip E. Johnson, *Darwin on Trial,* revised edition (Downer's Grove, IL: InterVarsity Press, 1993), 65.

20. Johnson, *Darwin on Trial,* 68, 151–52.

21. Richard C. Lewontin, "Billions and Billions of Demons," *New York Review of Books*, January 9, 1997.

22. "Brief of Amicus Curiae: 56 Professional Scientific Organizations in Support of Appellees and Affirmance," U.S. Court of Appeals for the Eleventh Circuit (June 9, 2005). Available online (June 2006) at: http://www. ncseweb.org/selman/SelmanScientistsBrief_final.pdf.

23. FASEB Board of Directors, "FASEB Opposes Using Science Classes to Teach Intelligent Design, Creationism, and other Non-Scientific Beliefs," Federation of American Societies for Experimental Biology, December 19,

2005. Available online (June 2006) at: http://opa.faseb.org/pdf/Evolution-Statement.pdf.

Chapter Seven:
You'd Think Darwin Created the Internet

1. Theodosius Dobzhansky, "Nothing in Biology Makes Sense Except in the Light of Evolution," *American Biology Teacher* 35 (March 1973), 125–29.

2. Bruce Alberts, "Preface," *Science and Creationism: A View from the National Academy of Sciences* (Washington, DC: National Academy Press, 1999). Available online (April 2006) at: http://books.nap.edu/html/creationism/.

3. Douglas J. Futuyma, *Evolution* (Sunderland, MA: Sinauer Associates, 2005), xiv.

4. John T. Schlebecker, *Whereby We Thrive: A History of American Farming, 1607–1972.* (Ames, IA: The Iowa State University press, 1975), 174–87, 316–18. Percy W. Bidwell and John I. Falconer, *History of Agriculture in the Northern United States, 1620–1860* (Washington, DC: Carnegie Institution, 1925; reprinted in 1973 by Augustus M. Kelley Publishers). "British Agricultural Revolution," Wikipedia: The Free Encyclopedia. Available online (April 2006) at: http://en.wikipedia.org/wiki/British_Agricultural_Revolution.

5. Nicholas Russell, *Like Eengend'ring Like: Heredity and Animal Breeding in Early Modern England* (Cambridge: Cambridge University Press, 1986), 39, 216. Charles Darwin, *The Variation of Animals and Plants Under Domestication* (New York: D. Appleton, 1868), Volume II, Chapter XX. Peter J. Bowler, *Evolution: The History of an Idea*, Revised Edition (Berkeley, CA: University of California Press, 1989), 155–56, 166.

6. William Bateson, *Mendel's Principles of Heredity* (New York: G. P. Putnam's Sons, 1913), 329. B. C. A. Windle, "Mendel, Mendelism," in *The Catholic Encyclopedia,* Volume X (Robert Appleton Company, 1911). Available online (April 2006) at: http://www.newadvent.org/cathen/10180b.htm.

7. Charles Darwin, *The Variation of Animals and Plants under Domestication*, Chapter XXVII. Charles Darwin, *On the Origin of Species*, Sixth Edition, Chapter V. See also Chapters I and VI. Bowler, *Evolution: The History of an Idea*, 171, 190, 210, 250–52.

8. Bateson, *Mendel's Principles of Heredity,* 334. Windle, "Mendel, Mendelism," in *The Catholic Encyclopedia.* Jan Sapp, *Beyond the Gene: Cytoplasmic Inheritance and the Struggle for Authority in Genetics.* (New York: Oxford University Press, 1987), Chapters 2–4.

9. Michael L. Dini, "Letters of Recommendation," Available online (April 2006) at http://www2.tltc.ttu.edu/dini/Personal/letters.htm.

10. Thomas McKeown, *The Role of Medicine* (Princeton: Princeton University Press, 1979). Kenneth F. Kiple, "The History of Disease," 16–50 in Roy Porter (editor), *The Cambridge Illustrated History of Medicine* (Cambridge: Cambridge University Press, 1996), 39–40.

11. Sherwin B. Nuland, *The Doctors' Plague: Germs, Childbed Fever, and the Strange Story of Ignác Semmelweis* (New York: W. W. Norton, 2003). Milton Wainwright, *Miracle Cure: The Story of Penicillin and the Golden Age of Antibiotics* (Oxford: Basil Blackwell, 1990), 11–12.

12. F. Fenner, D. A. Henderson, I. Arita, Z. Jezek, and I. D. Ladnyi, *Smallpox and its Eradication* (Geneva: World Health Organization, 1988). Abbas M. Behbehani, *The Smallpox Story* (Kansas City, KS: University of Kansas Medical Center, 1988).

13. Carl Zimmer, *Evolution: The Triumph of an Idea* (New York: Harper Collins, 2001), 336.

14. Alexander Fleming, "On the antibacterial action of cultures of a Penicillium, with special reference to their use in the isolation of *B. influenzae,*" *British Journal of Experimental Pathology* 10 (1929), 226–36. Wainwright, *Miracle Cure,* Chapter 2. Hare, Ronald, *The Birth of Penicillin* (London: George Allen and Unwin, 1970).

15. E. Chain, H. W. Florey, A. D. Gardner, N. G. Heatley, M. A. Jennings, J. Orr-Ewing, and A. G. Sanders, "Penicillin as a chemotherapeutic agent," *The Lancet* 239:2 (August, 1940), 226–28. Wainwright, *Miracle Cure,* Chapters 3 and 4. Alexander Fleming, "Banquet Speech," December 10, 1945. Available online (April 2006) at: http://nobelprize.org/medicine/laureates/1945/fleming-speech.html. Ronald W. Clark, *The Life of Ernst Chain: Penicillin and Beyond* (London: Weidenfeld and Nicolson, 1985), 147–48.

16. Albert Schatz, Elizabeth Bugie, and Selman A. Waksman, "Streptomycin, a Substance Exhibiting Antibiotic Activity Against Gram-Positive and Gram-Negative Bacteria," *Proceedings of the Society for Experimental*

Biology and Medicine 55 (1944), 66–69. Wainwright, *Miracle Cure,* Chapter 8. Selman A Waksman, "The Role of Antibiotics in Natural Processes," *Giornale di Microbiologia* 2 (1956), 1–14.

17. Joe J. Harrison, Raymond J. Turner, Lyriam L. R. Marques, and Howard Ceri, "Biofilms: A new understanding of these microbial communities is driving a revolution that may transform the science of microbiology," *American Scientist* 93 (November/December 2005): 508–15. Roland F. Hirsch, "Darwinian Evolutionary Theory and the Life Sciences in the Twenty-First Century," in William A. Dembski, Editor, *Uncommon Dissent: Intellectuals Who Find Darwinism Unconvincing* (Wilmington, DE: ISI Books, 2004), 215–31.

18. Alan R. Hinman, "Perspectives on Emergence and Control of Infectious Diseases Worldwide," 2–9 in Sarah S. Long, Larry K. Pickering, and Charles G. Prober (eds.), *Principles and Practice of Pediatric Infectious Diseases*, Second Edition (New York: Elsevier/Churchill Livingstone, 2003), 6. Larry J. Straus-baugh and Daniel B. Jernigan, "Emerging Infections," 107–16 in Sherwood L. Gorbach, John G. Bartlett, and Neil R. Blacklow (Editors), *Infectious Diseases*, Third Edition (Philadelphia, PA: Lippincott Williams & Wilkins, 2004), 111.

19. Steven M. Opal and Antone A. Medeiros, "Molecular Mechanisms of Antibiotic Resistance in Bacteria," 253–70 in Gerald L. Mandell, John E. Bennett, and Raphael Dolin (Editors), *Mandell, Douglas, and Bennett's Principles and Practice of Infectious Diseases*, Sixth Edition (Philadelphia, PA: Elsevier/Churchill Livingstone, 2005), 266.

20. Stewart T. Cole and Pedro M. Alzari, "TB—A New Target, a New Drug," *Science* 307 (2005), 214–15. Chaitan Khosla and Yi Tang, "A New Route to Designer Antibiotics," *Science* 308 (2005), 367–68. Malcolm Mac-Coss and Thomas A. Baillie, "Organic Chemistry in Drug Discovery," *Science* 303 (2004), 1810–13.

21. William C. Dampier, *A History of Science* (Cambridge: Cambridge University Press, 1966). See also the relevant entries in Wikipedia, The Free Encyclopedia, available online (April 2006) at: http://en.wikipedia.org/wiki/Main_Page.

22. *Getting the Facts Straight: A Viewer's Guide to PBS's Evolution* (Seattle, WA: Discovery Institute Press, 2001), Chapter 7. Available online (April 2006) at: http://www.reviewevolution.org/.

23. Kirschner is quoted in Peter Dizikes, "Missing Links," *Boston Globe*, October 23, 2005. Available online (April 2006) at: http://www.boston. com/news/globe/ideas/articles/2005/10/23/missing_links/. Philip S. Skell, "Why Do We Invoke Darwin?" *The Scientist* 19:16 (August 29, 2005): 10. Available online (April 2006) at: http://www.the-scientist.com/2005/8/29/10/1/.

24. Transcript: Vice President Gore on CNN's "Late Edition," March 9, 1999. Available online (April 2006) at: http://www.cnn.com/ALLPOLITICS/ stories/1999/03/09/president.2000/transcript.gore/. Declan McCullagh, "I Created the 'Al Gore Created the Internet' Story," *Politech*, October 17, 2000. Available online (April 2006) at: http://seclists.org/lists/politech/2000/Oct/0032.html.

25. Richard Wiggins, "Al Gore and the Creation of the Internet," *First Monday* 5:10 (October 2000). Available online (April 2006) at: http://www.firstmonday.dk/issues/issue5_10/wiggins/. John Markoff, "Sharing the Supercomputers," *New York Times*, December 29, 1988.

Chapter Eight:
The Design Revolution

1. William A. Dembski, *The Design Inference: Eliminating Chance Through Small Probabilities* (Cambridge: Cambridge University Press, 1998), 19, 36.

2. Ibid., 38.

3. Ibid., 3.

4. Ibid., 66. William A. Dembski, *The Design Revolution: Asking the toughest questions about intelligent design* (Downers Grove, IL: InterVarsity Press, 2004), 27. William A. Dembski, *No Free Lunch: Why Specified Complexity Cannot Be Purchased Without Intelligence* (Lanham, MD: Rowman & Littlefield, 2002).

5. Dembski, *The Design Revolution,* Chapter 6.

6. Richard Dawkins, *The Blind Watchmaker: Why the Evidence of Evolution Reveals a Universe Without Design* (New York: W. W. Norton, 1986), 47–49.

7. Ibid., 46, 50.

8. Timothy McGrew, "Toward a Rational Reconstruction of Design Inferences," *Philosophia Christi* 7 (2005), 253–98. Lydia McGrew, "Testability,

Likelihoods, and Design," *Philo* 7 (2004): 5–21. Available online (April 2006) at: http://www.lydiamcgrew.com/PhiloTestability.pdf. Paul Thagard, "Inference to the Best Explanation: Criteria for Theory Choice," *Journal of Philosophy* 75 (1978), 76–92. Peter Lipton, *Inference to the Best Explanation*, Second Edition (London: Routledge, 2004).

9. Fred Heeren, "The Lynching of Bill Dembski," *American Spectator* (November 2000), 44–50. Available online (April 2006) at: http://www.discovery.org/scripts/viewDB/index.php?command=view&program=CSCStories&id=532.

10. Forrest testimony, *Kitzmiller* v. *Dover Area School District* trial transcript, Day 6 (October 5), PM Session, Part 2. Available online (April 2006) at: http://www.talkorigins.org/faqs/dover/day6pm2.html. "Statement of Principles," New Orleans Secular Humanist Association. Available online (April 2006) at: http://nosha.secularhumanism.net/. Barbara Carroll Forrest, "Curriculum Vitae," *Internet Infidels*. Available online (April 2006) at: http://www.infidels.org/library/modern/barbara_forrest/bio.html. Barbara Forrest, "Letter to Invitees of the 'Nature of Nature' Conference," March 2000. Available online (April 2006) at: http://www.pandasthumb.org/archives/2005/10/the_pseudo-scie.html.

11. Program and Schedule for THE NATURE OF NATURE: An Interdisciplinary Conference on the Role of Naturalism in Science (April 12–15, 2000). Available online (April 2006) at: http://www.designinference.com/documents/2000.04.nature_of_nature.htm.

12. U.S. Representative Mark Souder (R-IN) in the Congressional Record, June 14, 2000. Available online (April 2006) at: http://www.baptist2baptist.net/b2barticle.asp?ID=71.

13. Blair Martin, "Professors Debate Legitimacy of Polanyi," *Baylor Lariat*, April 12, 2000. Available online (April 2006) at: http://www.baylor.edu/Lariat/news.php?action=story&story=15134. "Baylor Conference on Naturalism in Science Sparks Controversy," *Science & Theology News*, September 1, 2000. Available online (April 2006) at: http://www.stnews.org/News-2397.htm. Fred Heeren, "The Deed Is Done," *American Spectator* (December 2000–January 2001), 28–29.

14. "Infinite Monkey Theorem," Answers.com. Available online (April 2006) at: http://www.answers.com/topic/infinite-monkey-theorem.

15. The Monkey Shakespeare Simulator http://user.tninet.se/~ecf599g/aardasnails/java/Monkey/webpages/index.html.

16. David Adam, "Give Six Monkeys a Computer, and What Do You Get? Certainly Not the Bard," *Guardian*, May 9, 2003. "No Words to Describe Monkey's Play," BBC News, May 9, 2003. Available online (April 2006) at: http://news.bbc.co.uk/1/3013959.stm. Brian Bernbaum, "Monkey Theory Proven Wrong," CBS News: The Odd Truth, May 9, 2003. Available online (April 2006) at: http://www.cbsnews.com/stories/2003/05/12/national/main553500.shtml.

Chapter Nine:
The Secret of Life

1. Horace Freeland Judson, *The Eighth Day of Creation: The Makers of the Revolution in Biology* (New York: Simon and Schuster, 1979), 172–75.

2. Ibid., 216–17.

3. J. D. Watson and F. H. C. Crick, "A Structure for Deoxyribose Nucleic Acid," *Nature* 171 (1953): 737–38. J. D. Watson and F. H. C. Crick, "Genetical Implications of the Structure of Deoxyribonucleic Acid," *Nature* 171 (1953): 964–67.

4. Stephen C. Meyer, "DNA and Other Designs," *First Things* 102 (April 2000): 30–38. Available online (June 2006) at: http://www.discovery.org/scripts/viewDB/index.php?Command=view&id=200. Bill Gates, with Nathan Myhrvold and Peter Rinearson, *The Road Ahead* (New York: Penguin Books, 1995), 188.

5. Josh P. Roberts, "Making Do with the Bare Minimum," *The Scientist*, January 2006, 65–66. Eugene V. Koonin, "How Many Genes Can Make a Cell: The Minimal-Gene-Set Concept," *Annual Review of Genomics and Human Genetics* 1 (2000): 99–116.

6. Francis Crick, "On Protein Synthesis," *Symposia of the Society for Experimental Biology* 12 (1958): 138–63.

7. Charles B. Thaxton, Walter L. Bradley, and Roger L. Olsen, *The Mystery of Life's Origin* (Dallas, TX: Lewis and Stanley, 1984), 210–11.

8. Stephen C. Meyer, "DNA and the Origin of Life: Information, Specification, and Explanation," 223–85. John Angus Campbell and Stephen C. Meyer (editors), *Darwinism, Design, and Public Education* (East Lansing, MI: Michigan State University Press, 2003).

9. Meyer, "DNA and the Origin of Life." Meyer, "DNA and Other Designs."

10. Michael Polanyi, "Life's Irreducible Structure," *Science* 160 (1968), 1308–12.

11. Meyer, "DNA and the Origin of Life."

12. Kenneth R. Miller, "How Intelligent is Intelligent Design," *First Things* 106 (October 2000), 2–3. Stephen C. Meyer, "How Intelligent is Intelligent Design," *First Things* 106 (October 2000), 4–5. Available online (June 2006) at: http://www.firstthings.com/ftissues/ft0010/correspondence.html#intelligent.

13. Stephen C. Meyer, "The origin of biological information and the higher taxonomic categories," *Proceedings of the Biological Society of Washington* 117 (2004), 213–39.

14. Sternberg's account of the publication of Meyer's article is available online (January 2006) at: http://www.rsternberg.net/publication_details.htm. Meyer's article itself is available online (January 2006) at: http://www.discovery.org/scripts/viewDB/index.php?command=view&id=2177&program=CSC%20-%20Scientific%20Research%20and%20Scholarship%20-%20Science. Additional materials are also available online (January 2006) at: http://www.discovery.org/scripts/viewDB/index.php?command=view&program=CSC%20-%20Views%20and%20News&id=2399.

15. Discovery Institute Staff, "Peer-Reviewed, Peer-Edited, and other Scientific Publications Supporting the Theory of Intelligent Design (Annotated)," February 22, 2006. Available online (April 2006) at: http://www.discovery.org/scripts/viewDB/index.php?command=view&id=2640&program=CSC%20-%20Scientific%20Research%20and%20Scholarship%20-%20Science.

16. Michael J. Behe, "Correspondence with Science Journals: Response to Critics Concerning Peer-Review," Discovery Institute (August 2, 2000). Available online (April 2006) at: http://www.discovery.org/scripts/viewDB/index.php?command=view&id=450.

17. Jonathan Wells, "Catch-23," *Research News & Opportunities in Science and Theology,* July/August 2002. Available online (April 2006) at: http://www.discovery.org/scripts/viewDB/index.php?command=view&id=1212.

18. Jim Giles, "Peer-reviewed paper defends theory of intelligent design," *Nature* 431 (2004): 114. Trevor Stokes, "Intelligent design study appears,"

The Scientist 5 (September 3, 2004): 4. Available online (April 2006) at: http://www.the-scientist.com/news/20040903/04/ See the home page of *Richard M.* v. *Sternberg*, available online (April 2006) at: http://www.rsternberg.net/. The NCSE's and SI's misbehavior were documented by the U.S. Office of Special Counsel (OSC). The OSC's August 2005 report is available online (April 2006) at: http://www.rsternberg.net/OSC_ltr.htm.

19. David Klinghoffer, "The Branding of a Heretic," *Wall Street Journal*, January 28, 2005. Available online (January 2006) at: http://www.opinion-journal.com/taste/?id=110006220. Joyce Howard Price, "Researcher Claims Bias by Smithsonian," *Washington Times*, February 13, 2005. Available online (January 2006) at: http://www.washingtontimes.com/national/20050213-121441-8610r.htm. Letter from the U.S. Office of Special Counsel to Dr. Richard Sternberg, August 5, 2005. Available online (January 2006) at: http://www.rsternberg.net/Documents/OSC-Sternberg-preclosure-ltr2.pdf. Barbara Bradley Haggerty, "Profile: Intelligent Design and Academic Freedom," transcript of "All Things Considered" (NPR), November 10, 2005. Available online (January 2006) at: http://www.discovery.org/scripts/viewDB/index.php?command=view&program=CSC%20-%20Views%20and%20News&id=3083.

20. Richard Monastersky, "Biology Journal Says It Mistakenly Published paper That Attacks Darwinian Evolution," *Chronicle of Higher Education Daily News*, September 10, 2004. "Statement from the Council of the Biological Society of Washington," October 4, 2004. Available online (April 2006) at: http://www.biolsocwash.org/id_statement.html.

Chapter Ten:
Darwin's Black Box

1. "Ole Evinrude," *Inventor of the Week Archive*, Lemelson-MIT Program, January 1999. Available online (April 2006) at: http://web.mit.edu/invent/iow/evinrude.html. Kenneth Bjork, "Ole Evinrude and the Outboard Motor," The Norwegian-American Historical Association. Available online (April 2006) at: http://www.naha.stolaf.edu/pubs/nas/volume12/vol12_9.htm.

2. David J. DeRosier, "The Turn of the Screw: The Bacterial Flagellar Motor," *Cell* 93 (1998), 17–20.

3. Michael J. Katz, *Templets and the explanation of complex patterns* (Cambridge: Cambridge University Press, 1986), 26–27, 65–66, 83.

4. Charles Darwin, *The Origin of Species*, Sixth Edition, Chapter VI. Michael J. Behe, *Darwin's Black Box* Tenth Anniversary Edition (New York: The Free Press, 2006), 39.

5. John H. McDonald, "A reducibly complex mousetrap." Available online (April 2006) at: http://udel.edu/~mcdonald/mousetrap.html. Michael J. Behe, "Comments on Ken Miller's Reply to My Essays," January 8, 2001. Available online (April 2006) at http://www.discovery.org/scripts/viewDB/index.php?command=view&id=579.

6. Jerry A. Coyne, "The Case Against Intelligent Design: The Faith That Dare Not Speak Its Name," in *New Republic*, August 22, 2005.

7. Behe, *Darwin's Black Box,* 22.

8. Russell F. Doolittle, "A Delicate Balance," *Boston Review* (February/March 1997). Available online (April 2006) at: http://www.bostonreview.net/br22.1/doolittle.html.

9. Michael J. Behe, "In Defense of the Irreducibility of the Blood Clotting Cascade: Response to Russell Doolittle, Ken Miller and Keith Robison," July 31, 2000. Available online (April 2006) at: http://www.discovery.org/scripts/viewDB/index.php?command=view&id=442.

10. Robert M. Macnab, "Flagella," 70–83 in Frederick C. Neidhardt, et al. (editors), *Escherischia coli and Salmonella typhimurium: Cellular and Molecular Biology* (Washington, DC: American Society for Microbiology, 1987), Volume 1. Howard C. Berg, "The Rotary Motor of Bacterial Flagella" *Annual Review of Biochemistry* 72 (2003), 19–54.

11. Kenneth R. Miller, "The Flagellum Unspun: The Collapse of Irreducible Complexity," 81–97 in William A. Dembski and Michael Ruse (editors), *Debating Design: From Darwin to DNA* (Cambridge: Cambridge University Press, 2004). Available online (April 2006) at: http://www.millerandlevine.com/km/evol/design2/article.html.

12. Michael J. Behe, "Irreducible Complexity: Obstacle to Darwinian Evolution," 352–70 in William A. Dembski and Michael Ruse (editors), *Debating Design: From Darwin to DNA* (Cambridge: Cambridge University Press, 2004).

13. Scott A. Minnich and Stephen C. Meyer, "Genetic Analysis of Coordinate Flagellar and Type III Regulatory Circuits," 295–304 in M. W. Collins and C. A. Brebbia (editors), *Proceedings of the Second International Conference on Design & Nature* (Rhodes, Greece: WIT Press, 2004).

14. Milton H. Saier, Jr., "Evolution of Bacterial Type III Protein Secretion Systems," *Trends in Microbiology* 12 (2004), 113–15. Minnich and Meyer, "Genetic Analysis of Coordinate Flagellar and Type III Regulatory Circuits."

15. Michael Rubinkam, "Intelligent Design Proponent an Outcast at Own University," Associated Press, October 14, 2005.

16. University of Idaho president Timothy P. White, "Letter to the University of Idaho Faculty, Staff and Students," October 5, 2005. Available online (April 2006) at: http://www.president.uidaho.edu/default.aspx?pid=85947. David Epstein, "Drawing a Line in the Academic Sand," *Inside Higher Ed*, October 6, 2005. Available online (April 2006) at: http://www.insidehigh-ered.com/news/2005/10/06/idaho. "Discovery Institute Denounces University of Idaho's Ban on Differing Views on Evolution as Unconstitutional," October 4, 2005. Available online (April 2006) at: http://www.discovery.org/scripts/viewDB/index.php?command=view&id=2911&program=News&callingPage=discoMainPage.

17. "AAUP Statement on Professor Ward Churchill Controversy," American Association of University Professors, February 3, 2005. Available online (April 2006) at: http://www.aaup.org/newsroom/Newsitems/churchill.htm. John Miller, "U of I President: Teach Only Evolution in Science Classes," Associated Press, October 6, 2005. Available online (April 2006) at: http://www.discovery.org/scripts/viewDB/index.php?command=view&program=CSC%20%20Views%20and%20News&id=2922.

Chapter Eleven:
What a Wonderful World

1. Daniel Fischer, "India'95: An Eclipse Full of Surprises." Available online (April 2006) at: http://www.geocities.com/CapeCanaveral/5599/india_95_story.html.

2. Guillermo Gonzalez and Jay W. Richards, *The Privileged Planet: How Our Place in the Cosmos Is Designed for Discovery* (Washington, DC: Regnery Publishing, 2004), 3. For more information go to: http://www.privilegedplanet.com/.

3. John D. Barrow and Frank J. Tipler, *The Anthropic Cosmological Principle* (Oxford: Clarendon Press, 1986). Roger Penrose, *The Emperor's New Mind* (New York: Oxford University Press, 1989), 344.

4. Lawrence J. Henderson, *The Fitness of the Environment,* Reprint of the 1913 edition (Boston: Beacon Press, 1958). Michael J. Denton, *Nature's Destiny: How the Laws of Biology Reveal Purpose in the Universe* (New York: The Free Press, 1998).

5. Michael J. Denton, *Nature's Destiny:* 45, 108–09.

6. John Leslie, "Anthropic Principle, World Ensemble, Design," *American Philosophical Quarterly* 19, (1982), 141–50.

7. Stephen C. Meyer, "Evidence for Design in Physics and Biology," Wethersfield Institute, 53–111. *Science and Evidence for Design in the Universe* (San Francisco: Ignatius Press, 2000), 58–62. Gonzalez and Richards, *The Privileged Planet,* 265–71.

8. Guillermo Gonzalez, Donald Brownlee, and Peter D. Ward, "Refuges for Life in a Hostile Universe," *Scientific American* (October 2001), 60–67. Peter D. Ward and Donald Brownlee, *Rare Earth: Why Complex Life Is Uncommon in the Universe* (New York: Springer-Verlag, 2000).

9. Gonzalez and Richards, *The Privileged Planet,* 6–7, 55–60, 127–36.

10. Ibid., xv, 327.

11. Illustra Media, "The Privileged Planet" (La Habra, CA: Illustra Media, 2004). For more information go to: http://www.illustramedia.com/tppinfo. htm.

12. Denyse O'Leary, "Design Film Sparks Angst: Under Fire, Smithsonian Disavows Presentation on Intelligent Design," *Christianity Today* (July 6, 2005). Available online (April 2006) at: http://www.christianitytoday. com/ct/2005/008/4.22.html.

13. John Schwartz, "Smithsonian to Screen a Movie That Makes a Case Against Evolution," *New York Times*, May 28, 2005. Available online (April 2006) at: http://www.nytimes.com/2005/05/28/national/28smithsonian. html?ex=1142139600&en=cc3a2302008f868f&ei=5070.

14. Trevor Stokes, "... And Smithsonian Has ID Troubles," *The Scientist* 19 (July 4, 2005): 13. Available online (April 2006) at: http://www.the-scientist.com/2005/07/04/13/1/. "Museum Quits as Film Sponsor," *New York Times*, June 3, 2005. Available online (April 2006) at: http://www.nytimes. com/2005/06/03/national/03smithsonian.html?ex=1141966800&en=36eda8c 0892f6b07&ei=5070.

15. "Smithsonian Backs Off Intelligent Design Film: Museum Pulls Sponsorship After Darwinists Pressure," *WorldNetDaily,* June 4, 2005. Available

online (April 2006) at: http://worldnetdaily.com/news/article.asp?ARTI-CLE_ID=44599.

16. Discovery Institute Staff, "*The Privileged Planet* National Premiere at the Smithsonian Institution's National Museum of Natural History," June 23, 2005. Available online at: http://www.discovery.org/scripts/viewDB/index.php?command=view&id=2703.

17. James Randi, "Important Notice," *Online Newsletter of the James Randi Educational Foundation*, May 27, 2005. Available online (April 2006) at: http://www.randi.org/jr/052705a.html. David Berlinski, "An Open Letter to the Amazing Randi," *Discovery Institute News* June 13, 2005. Available online (April 2006) at: http://www.discovery.org/scripts/viewDB/index.php?command=view&id=2670.

18. Letter, "Intelligent Design Not Supported by Science," *Iowa State Daily*, August 23, 2005. Available online (April 2006) at: http://www.iowastatedaily.com/media/paper818/news/2005/08/23/Opinion/Letter.Intelligent.Design.Not.Supported.By.Science-1104943.shtml?norewrite&sourcedomain=www.iowastatedaily.com.

19. Kate Strickler, "Intelligent Design theory sparks debate on campus," *Iowa State Daily*, August 25, 2005. Available online (April 2006) at: http://www.iowastatedaily.com/media/paper818/news/2005/08/25/News/Intelligent.Design.Theory.Sparks.Debate.On.Campus-1105025.shtml?norewrite&sourcedomain=www.iowastatedaily.com. Jamie Schumann, "120 Professors at Iowa State U. Sign Statement Criticizing Intelligent-Design Theory," *Chronicle of Higher Education*, August 25, 2005. Todd Dvorak, "ISU Professor Caught in ID Turmoil," *Hawk Eye*, February 11, 2006. Available online (April 2006) at: http://archive.thehawkeye.com/.

20. Reid Forgrave, "A Universal Debate: An ISU Astronomy Professor Finds Himself at the Center of a Controversy over Science and Religion," *Des Moines Register*, August 31, 2005. Available online (April 2006) at: http://www.dmregister.com/apps/pbcs.dll/article?AID=/20050831/LIFE/%20508310325/1001/LIFE. Guillermo Gonzalez, "ID Theory: Open to 'Design' Possibilities," *Des Moines Register*, August 31, 2005. Available online (April 2006) at: http://www.discovery.org/scripts/viewDB/index.php?command=view&id=2824.

21. Forgrave, "A Universal Debate." Kate Strickler, "U of I Joins Regent Institutions in Intelligent Design Discussion," *Iowa State Daily*, October 19, 2005. Available online (April 2006) at: http://www.iowastatedaily.com/media/

paper818/news/2005/10/19/News/U.Of-I.Joins.Regent.Institutions.In.Intelli-gent.Design.Discussion-1106604.shtml?norewrite&sourcedomain=www. iowastatedaily.com. Frederick Skiff, Personal communication, November 18, 2005.

22. Dave Schweingruber, "A Witch Hunt at ISU," *Mid-Iowa News*, August 30, 2005. Available online (April 2006) at: http://www.midiowanews.com/site/news.cfm?newsid=15121481&BRD=27 00&PAG=461&dept_id=554188&rfi=8.

23. Iowa State University Atheist and Agnostic Society, http://www.stuorg. iastate.edu/isuaas. Avalos was a featured speaker at the Atheist Alliance con-ference in Kansas City, MO, April 14–16, 2006: http://atheistalliance.org/con-ventions/2006/index.php.

Chapter Twelve:
Is ID Science?

1. Letter from ASCB president Harvey F. Lodish to Ohio governor Bob Taft, February 24, 2004. Available online (June 2006) at: http://www. ascb.org/publicpolicy/taft04.htm. Statement on the Teaching of Evolu-tion, American Astronomical Society, September 20, 2005. Available online (June 2006) at: http://www.aas.org/governance/council/resolutions. html#teach.

2. Statement on Teaching Alternatives to Evolution, Biophysics Society, November, 2005. Available online (June 2006) at: http://www.biophysics.org/ pubaffairs/issues.htm. John Thavis, "Intelligent Design Not Science, Says Vatican Newspaper Article," Catholic News Service, January 17, 2006. Available online (June 2006) at: http://www.catholicnews.com/data/sto-ries/cns/0600273.htm.

3. Judge William R. Overton, "United States District Court Opinion *McLean* v. *Arkansas*," 307–31 in Michael Ruse (editor), *But Is It Science?* (Amherst, NY: Prometheus Books, 1996).

4. Larry Laudan, "Science at the Bar—Causes for Concern," 351–55 in Michael Ruse (editor), *But Is It Science?* 353–54. Stephen C. Meyer, "The Methodological Equivalence of Design & Descent," 67–112 in J. P. Moreland (editor), *The Creation Hypothesis* (Downer's Grove, IL: InterVarsity Press, 1994). Available online (June 2006) at: http://www.discovery.org/ scripts/viewDB/index.php?command=view&id=1696.

5. Larry Laudan, "The Demise of the Demarcation Problem," 337–50 in Michael Ruse (editor), *But Is It Science?*

6. Michael Ruse, "Witness Testimony Sheet, *McLean* v. *Arkansas*," 287–306 in Michael Ruse (editor), *But Is It Science?*

7. Alvin Plantinga, "Whether ID Is Science Isn't Semantics," *Science & Theology News*, March 7, 2006. Available online (April 2006) at: http://www.stnews.org/Commentary-2690.htm.

8. Phillip E. Johnson, *Darwin on Trial*, Revised Edition (Downer's Grove, IL: InterVarsity Press, 1993), 117–18.

9. Del Ratzsch, "Design Theory and its Critics," *Ars Disputandi* 2 (October 28, 2002). Available online (June 2006) at: http://www.arsdisputandi. org/publish/articles/000079/article.pdf. Del Ratzsch, *Nature, Design, and Science: The Status of Design in Natural Science* (Albany, NY: State University of New York Press, 2001).

10. "Peer-Reviewed & Peer-Edited Scientific Publications Supporting the Theory of Intelligent Design (Annotated)," *Discovery Institute*, February 22, 2006. Available online (June 2006) at: http://www.discovery.org/scripts/ viewDB/index.php?command=view&id=2640&program=CSC%20-%20Scientific%20Research%20and%20Scholarship%20-%20Science.

11. "Leading Atheist Philosopher Concludes God's Real," FOX News (December 9, 2004). Available online (June 2006) at: http://www.foxnews. com/story/0,2933,141061,00.html. Stephen C. Meyer, "Intelligent Design Is Not Creationism," *Telegraph*, January 28, 2006. Available online (June 2006) at: http://www.telegraph.co.uk/opinion/main.jhtml?xml=/opinion/2006/01/28/do2803.xm. Richard C. Lewontin, "Billions and Billions of Demons," *New York Review of Books*, January 9, 1997. "Statement of Principles," New Orleans Secular Humanist Association. Available online (April 2006) at: http://nosha.secularhumanism.net/. Barbara Carroll Forrest, "Curriculum Vitae," *Internet Infidels*. Available online (April 2006) at: http://www.infidels.org/library/modern/barbara_forrest/bio.html.

12. Daniel C. Dennett, *Darwin's Dangerous Idea* (New York: Simon & Schuster, 1995), 18, 63, 310, 520. Richard Dawkins, *The Blind Watchmaker* (New York: W. W. Norton, 1986), 6. William B. Provine, "Darwin Day" address (Knoxville, TN: University of Tennessee, 1998). Available online (June 2006) at: http://members.iinet.net.au/~sejones/religi05.html# rlgnthsmvltnsngnf.

13. Larry Laudan, "More on Creationism," 363–66 in Michael Ruse (editor), *But Is It Science?*

14. Carl F. von Weizsächer, *The Relevance of Science* (1964), 151–53.

15. Larry Laudan, "The Demise of the Demarcation Problem," 338.

16. Alvin Plantinga, "Whether ID Is Science Isn't Semantics." Alvin Plantinga, "Methodological Naturalism?—Part II," *Origins & Design* 18 (1997), 22–34. Available online (June 2006) at: http://www.arn.org/docs/odesign/od182/methnat182.htm.

17. Larry Laudan, "The Demise of the Demarcation Problem," 349. Larry Laudan, "Science at the Bar—Causes for Concern," 354–55.

Chapter Thirteen:
To Teach, or Not to Teach

1. *Skagit Valley Herald*, May 30 1998. Available online (June 2006) at: http://www.skagitvalleyherald.com/articles/1998/05/30/news21096.txt.

2. *Skagit Valley Herald*, May 16 2001, A1, and Editorial for July 29, 2001. Available online (June 2006) at: http://www.skagitvalleyherald.com/articles/2001/05/16/news22854.txt and http://www.skagitvalleyherald.com/articles/2001/07/29/news25198.txt. See also "Unscientific Methods," *World* magazine, September 18, 2004. Available online (June 2006) at: http://www.worldmag.com/subscriber/displayarticle.cfm?id=9637.

3. Doug Cowan, "Teaching Students to be 'Competent Jurors' on Evolution," *Christian Science Monitor*, May 31, 2005. Available online (June 2006) at: http://www.csmonitor.com/2005/0531/p09s01-coop.html. Doug Cowan, "Foster Critical Thinking in Schools," *Seattle Post-Intelligencer,* October 10, 2005. Available online (June 2006) at: http://seattlepi.nwsource.com/opinion/243931_firstperson09.html.

4. "No Child Left Behind Act of 2001: Conference Report," *U.S. House of Representatives Report 107-334* (Washington, DC: U.S. Government Printing Office, 2001), 703. Available online (June 2006) at: http://www.discovery.org/scripts/viewDB/filesDB-download.php?id=113.

5. Kenneth R. Miller, "The Truth about the 'Santorum Amendment' Language on Evolution." Available online (June 2006) at: http://www.millerandlevine.com/km/evol/santorum.html. Letter to Bruce Chapman (Discovery Institute) from U.S. Representative John A. Boehner (R-OH), U.S. Senator Judd Gregg (R-NH), and U.S. Senator Rick Santorum (R-PA)

regarding the significance of the Conference Report, September 10, 2003. Available online (June 2006) at: http://www.discovery.org/scripts/viewDB/ filesDB-download.php?id=112. "No Child Left Behind Act and the 'Santorum Language'," Discovery Institute (July 1, 2004). Available online (June 2006) at: http://www.discovery.org/scripts/viewDB/index.php?command= view&id=2103.

6. Eugenie C. Scott, "Not (Just) in Kansas Anymore," *Science* 288 (2000): 813–15. Eugenie C. Scott and Glenn Branch, "Evolution: What's Wrong with 'Teaching the Controversy'?" *Trends in Ecology and Evolution* 18 (2003), 499–502.

7. Stephen C. Meyer, "Teaching about Scientific Dissent from Neo-Darwinism," *Trends in Ecology and Evolution* 19 (2004), 115–16. Eugenie C. Scott and Glenn Branch, "Teaching the Controversy: Response to Langen and Meyer," *Trends in Ecology and Evolution* 19 (2004), 116–17.

8. Douglas J. Futuyma, *Evolutionary Biology*, Third Edition (Sunderland, MA: Sinauer Associates, 1998), 759–60. Peter H. Raven & George B. Johnson, *Biology*, Sixth Edition (Boston, MA: McGraw-Hill, 2002), 455–56. Sylvia S. Mader, *Biology*, Eighth Edition (Boston, MA: McGraw-Hill, 2004), 300.

9. Douglas J. Futuyma, *Evolution* (Sunderland, MA: Sinauer Associates, 2005), 527, 534.

10. Steven D. Verhey, "The Effect of Engaging Prior Learning on Student Attitudes Toward Creationism and Evolution," *BioScience* 55 (2005), 996–1003. Jonathan Wells, *Icons of Evolution: Why Much of What We Teach About Evolution Is Wrong* (Washington, DC: Regnery Publishing, 2002).

11. "Kansas Inherits the Wind," *Washington Times National Weekly Edition*, August 23–29, 1999, 36. Jonathan Wells, "Ridiculing Kansas School Board Easy, but It's Not Good Journalism," *Daily Republic*, October 14, 1999. Available online (June 2006) at: http://www.discovery.org/scripts/ viewDB/index.php?command=view&id=63.

12. Rex Dalton, "Kansas Kicks Evolution Out of the Classroom," *Nature* 400, 1999, 701. Stephen Jay Gould, "Dorothy, It's Really Oz," *Time*, August 23, 1999, 59. Herbert Lin, "Kansas Evolution Ruling," *Science* 285, 1999, 1849. John Rennie, "A Total Eclipse of Reason," *Scientific American* 281, October, 1999, 124. John Rennie, "Fan Mail from the Fringe," *Scientific American*, February, 2000, 4.

13. Catherine Candisky, "Science Standards Proposal for Teaching Evolution in Schools Draws More Fire," *Columbus Dispatch*, January 15, 2002. Francis X. Clines, "In Ohio School Hearing, a New Theory Will Seek a Place Alongside Evolution," *New York Times*, February 11, 2002. William Saletan, "Unintelligent Redesign," *Slate*, February 13, 2002. Available online (June 2006) at: http://www.slate.com/id/2062009/.

14. Francis X. Clines, "Ohio Board Hears Debate on an Alternative to Darwinism," *New York Times*, March 12, 2002. "Evolution Debate Back in Ohio Schools," CNN Student News, March 12, 2002. Available online (June 2006) at: http://cnnstudentnews.cnn.com/2002/fyi/teachers.ednews/03/12/ohio.evolution.debate.ap/index.html.

15. "Ohio Strengthens Teaching of Evolution," *New York Times*, December 12, 2002.

16. Eileen Foley, "The Fanatics Don't Get What Science Is All About," *The Blade,* November 5, 2002. Catherine Candisky, "State OKs Curriculum Involving Creationism," *Columbus Dispatch*, March 10, 2004. Sam Fulwood, "State Board Creating Path for Creationism," *Plain Dealer*, February 7, 2004.

17. Liz Craig, Discussion Board, Kansas Citizens for Science, February 10, 2005. Available online (June 2006) at: http://www.kcfs.org/cgi-bin/ultimatebb.cgi?ubb=get_topic;f=3;t=000017.

18. Geoff Brumfiel, "Biologists Snub 'Kangaroo Court' for Darwin," *Nature* 434 (2005), 550. Edward W. Lempinen and Ginger Pinholster, "AAAS 'Respectfully declines' invitation to controversial Kansas evolution hearing," *EurekAlert* (April 12, 2005). Available online (June 2006) at: http://www.eurekalert.org/pub_releases/2005-04/plos-ad041205.php. Alan Leshner, "Letter to Kansas State Department of Education," American Association for the Advancement of Science, April 11, 2005. Available online (June 2006) at: http://www.eurekalert.org/images/release_graphics/pdf/GriffithLetter.pdf.

19. Barbara Hollingsworth, "Science Faces Trial: State School Board Panel Begins Hearings Today," *Topeka Capital-Journal*, May 5, 2005.

20. "Kansas Curricular Standards for Science Education," *Kansas State Department of Education,* November 8, 2005. Available online (June 2006) at: http://www.ksde.org/outcomes/sciencestd.html. Jonathan Wells, "Definitions of State Science Standards," Discovery Institute, November 10, 2005.

Available online (June 2006) at: http://www.discovery.org/scripts/viewDB/ index.php?command=view&id=2573.

21. Edward Sisson, "Darwin Takes the Fifth: What Really Happened at the Kansas Evolution Hearings," *Touchstone*, September 2005. Available online (June 2006) at: http://touchstonemag.com/archives/article.php?id=18-07-054-r. "Science Standards Expert Testimony," Kansas State Department of Education, May 5–12, 2005. Available online (June 2006) at: http://www.ksde.org/outcomes/sciencestdexptest.html.

22. Edward W. Lempinen, "AAAS 'Deeply Disturbed' by Kansas State Board of Education Vote, Leshner Says," American Association for the Advancement of Science, November 9, 2005. Available online (June 2006) at: http://www.aaas.org/news/releases/2005/1109kansas.shtml.

23. Jodi Rudoren, "Ohio Board Undoes Stand on Evolution," *New York Times*, February 15, 2006. "Darwinists Bully Ohio School Board into Censoring Teaching of Evolution," *Discovery Institute News* (February 14, 2006). Available online (June 2006) at: http://www.discovery.org/scripts/ viewDB/index.php?command=view&id=3257. Mark Bergin, "Junk science: Ohio school board caves to legal threats, dumps critical teaching on evolution," *World* magazine, February 25, 2006, 24. Available online (June 2006) at: http://www.worldmag.com/articles/11553. Martha Wise, "Conservative Ohio values led to change in evolution policy," *Cincinnati Enquirer* (February 22, 2006). Available online (June 2006) at: http://news.enquirer.com/ apps/pbcs.dll/article?AID=/20060222/EDIT02/602220305/-1/all.

24. Dover Policy, Memorandum Opinion (December 2005), 2. Available online (January 2006) at: http://www.pamd.uscourts.gov/kitzmiller/ kitzmiller_342.pdf.

25. "Discovery Calls Dover Evolution Policy Misguided, Calls for its Withdrawal," *Discovery Institute News*, December 14, 2004. Available online (June 2006) at: http://www.discovery.org/scripts/viewDB/index.php?command=view&id=2341. "Setting the Record Straight about Discovery Institute's Role in the Dover School District Case," *Discovery Institute News* (November 10, 2005). Available online (June 2006) at: http://www.discovery.org/scripts/viewDB/index.php?command=view&id=3003. "Discovery Institute's Science Education Policy," *Discovery Institute News*, January 16, 2006. Available online (June 2006) at: http://www.discovery.org/scripts/ viewDB/index.php?command=view&id=3164.

26. Judge Jones's decision is available online (January 2006) at: http://www.pamd.uscourts.gov/kitzmiller/kitzmiller_342.pdf. John G. West, "Dover in Review," The Discovery Institute, January 6, 2006. Available online (January 2006) at: http://www.discovery.org/scripts/viewDB/index.php?command=view&id=3135&program=CSC%20-%20Science%20and%20Education%20Policy%20-%20News%20and%20Articles. David K. DeWolf, John G. West, Casey Luskin and Jonathan Witt, *Traipsing Into Evolution: Intelligent Design and the Kitzmiller vs. Dover Decision* (Seattle, WA: Discovery Institute Press, 2006).

Chapter Fourteen:
Darwinism and Conservatives

1. Michael R. Bloomberg, "Address to Graduates of Johns Hopkins University School of Medicine," NYC.gov: The Official New York City Web Site, May 25, 2006. Available online (June 2006) at: http://www.nyc.gov/portal/index.jsp?epi_menuItemID=c0935b9a57bb4ef.

2. Greg Wilson, "Mike throws left at foes of evolution," *New York Daily News,* May 26, 2006. Available online (June 2006) at: http://www.nydailynews.com/news/politics/story/421069p-355478c.html.

3. Dan Froomkin, "Bush Backs Rove, Palmeiro, 'Intelligent Design'," *Washington Post,* August 2, 2005. Available online (June 2006) at: http://www.washingtonpost.com/wp-dyn/content/blog/2005/08/02/BL2005080201070_pf.html. Elisabeth Bumiller, "Bush Remarks Roil Debate Over Teaching of Evolution," *New York Times*, August 3, 2005. Mark Bergin, "Mad Scientists," *World Magazine*, August 20, 2005. Available online (June 2006) at: http://www.discovery.org/scripts/viewDB/index.php?command=view&id=2789.

4. John Derbyshire, " Teaching Science: The President Is Wrong on Intelligent Design," *National Review*, August 30, 2005. Available online (June 2006) at: http://www.nationalreview.com/derbyshire/derbyshire200508300823.asp.

5. Charles Krauthammer, "Phony Theory, False Conflict: 'Intelligent Design' Foolishly Pits Evolution Against Faith," *Washington Post*, November 18, 2005. Available online (June 2006) at: http://www.washingtonpost.com/wpdyn/content/article/2005/11/17/AR20051117 01304_pf.html.

6. George F. Will, "Grand Old Spenders," *Washington Post*, November 17, 2005. Available online (June 2006) at: http://www.washingtonpost.com/

wpdyn/content/article/2005/11/16/AR200511160 1883_pf.html. Allan H. Ryskind, "Darwinist Ideologues Are on the Run," *Human Events*, January 30, 2006. Available online (June 2006) at: http://www.humaneventson-line.com/article.php?id=11965. Casey Luskin, Future of Conservatism: Darwin or Design?" *Human Events,* December 12, 2005. Available online (June 2006) at: http://www.humaneventsonline.com/article.php?id=10790.

7. Larry Arnhart, *Darwinian Conservatism* (Charlottesville, VA: Imprint Academic, 2005), 143. James Seaton, "Natural Selection: Yet Another Reason to Admire the Author of 'The Origin of Species'," a review of Larry Arnhart's *Darwinian Conservatism, Weekly Standard*, May 8, 2006. Available online (June 2006) at: http://weeklystandard.com/Content/Public/Articles/000/000/012/157vbnww.asp.

8. Arnhart, *Darwinian Conservatism*, 1–3.

9. Edward O. Wilson, *Sociobiology* (Cambridge, MA: Harvard University Press, 1975), 382. The definition of altruism is compiled from the abridged paperback edition (1980), 55 and 312.

10. Stephen Jay Gould, "Sociobiology: the Art of Storytelling," *New Scientist*, November 16, 1978, 530–33. Jerry A. Coyne, "Of Vice and Men: The Fairy Tales of Evolutionary Psychology," a review of Randy Thornhill and Craig Palmer's *A Natural History of Rape*, in *New Republic*, April 3, 2000. Tom Bethell, "Against Sociobiology," *First Things* 109, January 2001: 18–24. Available online (June 2006) at: http://www.firstthings.com/ftissues/ft0101/articles/bethell.html.

11. Paul Lawrence Farber, *The Temptations of Evolutionary Ethics* (Berkeley: University of California Press, 1994), 2–5, 172–75. Carson Holloway, *The Right Darwin? Evolution, Religion, and the Future of Democracy* (Dallas, TX: Spence Publishing, 2006). Kate Campaigne, "Can New Darwinism Sustain a Healthy Polity?" American Enterprise Institute, May 4, 2006. Available online (June 2006) at: http://www.taemag.com/issues/articleID.19160/article_detail.asp. Amy Doolittle, "Survival of the Moralist," *Washington Times*, March 22, 2006. Available online (June 2006) at: http://washingtontimes.com/culture/20060322-120641-2603r.htm.

12. Charles Darwin, *The Descent of Man*, Chapter V.

13. Weikart, *From Darwin to Hitler,* 147. Cyril B. Means, Jr., "Eugenic Abortion," Letter to the Editor, *New York Times,* April 16, 1965. Letter from Milan

M. Vuitch to Senator John East, April 22, 1981, in The Human Life Bill Appendix: Hearings Before the Subcommittee on Separation of Powers of the Committee on the Judiciary, United States Senate, Ninety-seventh Congress, first session, on S. 158, a Bill to Provide that Human Life Shall be Deemed to Exist from Conception, April 23, 24; May 20, 21; June 1, 10, 12, and 18. Serial No. J-97-16 (Washington, DC: U.S. Government Printing Office, 1982), 105.

14. Charles Darwin, *The Descent of Man*, Chapter VI. Ernst Haeckel, *Anthropogenie oder Entwicklungsgeschichte des Menschen* (Leipzig: Verlag von Wilhelm Englemann, 1877), Figure XIV.

15. "Ota Benga," Wikipedia. Available online (June 2006) at: http://en. wikipedia.org/wiki/Ota_Benga. "Ota Benga Commits Suicide!" *African American Registry*, March 20, 1916. Available online (June 2006) at: http://www.aaregistry.com/african_american_history/1779/Ota_Benga_commits_suicide.

16. Charles Darwin, *The Origin of Species*, Chapter III. Richard Weikart, "Laissez-Faire Social Darwinism and Individualist Competition in Darwin and Huxley," *The European Legacy* 3, 1998, 17–30.

17. Richard Hofstadter, *Social Darwinism in American Thought*, Revised Edition (Boston: Beacon Press, 1955). Irvin G. Wyllie, "Social Darwinism and the Businessman," *Proceedings of the American Philosophical Society* 103 (October 1959).

18. Adam Smith, *An Inquiry into the Nature and Causes of the Wealth of Nations* (1776). Available online (June 2006) at: http://www.econlib.org/library/Smith/smWNtoc.html. Ludwig von Mises, *Socialism: An Economic and Sociological Analysis*, translated by J. Kahane (Indianapolis, IN: Liberty Fund, 1981), 281–86. Friedrich A. Hayek, "Freedom, Reason and Tradition," *Ethics* 68, July 1958. John G. West, "Darwin's Public Policy: Nineteenth Century Science and the Rise of the American Welfare State," John Marini and Ken Masugi (editors), *The Progressive Revolution in Politics and Political Science: Transforming the American Regime* (Lanham, Maryland: Rowman and Littlefield, 2005), 253–86.

19. George Gilder, *Wealth and Poverty* (New York: Basic Books, 1981), Chapters 19 and 20.

20. Ibid., 261–67.

21. George Gilder, "The Materialist Superstition," Discovery Institute, October 18, 2004. Available online (June 2006) at: http://www.discovery.org/

scripts/viewDB/index.php?command=view&id=2258. A heavily edited version of the essay was published in *Wired*, October 18, 2004. Available online (June 2006) at: http://www.wired.com/wired/archive/12.10/evolution.html?pg=5&topic=ev olution&topic_set=.

22. Gary Galles, "Is There Proof of Government's Intelligent Design?" Ludwig von Mises Institute Blog, October 04, 2005. Available online (June 2006) at: http://blog.mises.org/archives/004167.asp.

Chapter Fifteen:
Darwinism's War on Traditional Christianity

1. Hunter R. Rawlings III, State of the University Address, Cornell University Office of the President, October 21, 2005. Available online (June 2006) at: http://www.cornell.edu/president/announcement_2005_1021.cfm. Michelle York, "Cornell President Condemns Teaching Intelligent Design as Science," *New York Times*, October 22, 2005.

2. Andrew Dickson White, *A History of the Warfare of Science with Theology in Christendom* (New York: D. Appleton, 1896), v–ix, 1–88.

3. James R. Moore, *The Post-Darwinian Controversies* (Cambridge: Cambridge University Press, 1979). Tom Bethell, *The Politically Incorrect Guide to Science* (Washington, DC: Regnery Publishing, 2005), Chapter 12.

4. CIA-The World Factbook. Available online at: http://www.cia.gov/cia/ publications/factbook/geos/xx.html#People.

5. The Nicene Creed (325 A.D.). Available online (June 2006) at: http://www.creeds.net/ancient/nicene.htm. The Definition of the Council of Chalcedon (451 A.D.). Available online (June 2006) at: http://www. reformed.org/documents/index.html?mainframe=http://www.reformed.org/ documents/chalcedon.html.

6. Athanasius, "Discourses Against the Arians," "On the Incarnation of the Word," and "Against the Heathen," *Select Works and Letters* in Schaff (editor), *Nicene and Post-Nicene Fathers*. Augustine, "Two Souls: Against the Manichaeans," *Writings Against the Manichaeans and Against the Donatists* in Philip Schaff (editor), *Nicene and Post-Nicene Fathers* (Grand Rapids, MI: Eerdmans, 1974). Augustine, *Eighty-three Different Questions*, translated by David Mosher (Washington, DC: Catholic University Press, 1982), 46. Augustine, *On the Holy Trinity* and *Homilies on the Gospel of John*, in Schaff (edi-

tor), *Nicene and Post-Nicene Fathers*. Jonathan Wells, "Darwinism and the Argument to Design," *Dialogue & Alliance* 4 (January 1991): 69–85. Available online (June 2006) at: http://www.discovery.org/scripts/viewDB/index.php?command=view&id=102.

7. George Florovsky, "Idea of Creation in Christian Philosophy," *Eastern Churches Quarterly* 8 (1949-1950), 53–77. John Meyendorff, *Byzantine Theology* (New York: Fordham, 1974), 131–34. John Meyendorff, *Christ in Eastern Christian Thought* (Washington: Corpus, 1969), 100–01. Thomas Aquinas, *Summa Theologiae*, translated by Fathers of the English Dominican Province (New York: Benziger, 1947), Ia, q.15, art. 1 & 2; q. 14, art. 9; q. 19, art. 4; q. 34, art. 3; q. 44, art. 3. Thomas Aquinas, *Summa Contra Gentiles*, translated by Anton Pegis and James Anderson (London: Notre Dame, 1955-1956), book 1, chapter 72, art. 6. Additional references are cited in Wells, "Darwinism and the Argument to Design." Available online (June 2006) at: http://www.discovery.org/scripts/viewDB/index.php?command=view&id=102.

8. Martin Luther, *Works* (St. Louis: Concordia, 1958), 1:36–56, 80–84; 13:91–92; 17:29, 118; 22:8–12, 26–29. Paul Althaus, *Theology of Martin Luther* (Philadelphia: Fortress, 1966). John Calvin, *Institutes of the Christian Religion*, edited by John McNeill (Philadelphia: Westminster, 1960), book 1, chapters 5–6, 14, 16–18. Additional references are cited in Wells, "Darwinism and the Argument to Design." Jonathan Wells, *Charles Hodge's Critique of Darwinism: An Historical-Critical Analysis of Concepts Basic to the 19th Century Debate* (Lewiston, NY: Edwin Mellen Press, 1988).

9. Francis Darwin (editor), *The Life and Letters of Charles Darwin* (New York: D. Appleton, 1887), Volume I, 282–285; Volume II, 7, 97, 174, 217. Charles Darwin, *The Origin of Species*, Second through Sixth Editions, last sentence. Charles Darwin, *The Variation of Animals and Plants Under Domestication* (New York: Orange Judd, 1868), Volume II, 515–16.

10. Francis Darwin and A. C. Seward (editors), *More Letters of Charles Darwin* (New York: D. Appleton, 1903), Volume I, 321. F. Darwin (editor), *The Life and Letters of Charles Darwin*, Volume I, 278–79; Volume II, 105–06. Jonathan Wells, "Charles Darwin on the Teleology of Evolution," *International Journal on the Unity of the Sciences* 4 (Summer, 1991): 133–56.

11. George Gaylord Simpson, *The Meaning of Evolution*, Revised Edition (New Haven, CT: Yale University Press, 1967), 345. Monod was quoted in

Horace Freeland Judson, *The Eighth Day of Creation: The Makers of the Revolution in Biology* (New York: Simon and Schuster, 1979), 216–17. Stephen Jay Gould, *Ever Since Darwin* (New York: W. W. Norton, 1977), 147.

12. Richard Dawkins, "Is Science a Religion?" *The Humanist* (January/February 1997). Available online (June 2006) at: http://www.the-humanist.org/humanist/articles/dawkins.html. Daniel C. Dennett, *Darwin's Dangerous Idea* (New York: Simon & Schuster, 1995), 520. See advertisement in *Seed* magazine (December/January 2006), 84.

13. Laura Bauer, "Intelligent Design Backers Criticize KU Course Title," *Kansas City Star*, November 23, 2005. Available online (January 2006) at: http://www.kansascity.com/mld/kansascity/news/local/13237625.htm. Scott Jaschik, "Emails Kill a Course," *Inside Higher Education*, December 2, 2005. Available online (January 2006) at: http://insidehighered.com/news/2005/12/02/kansas. Michelle Malkin, "What Happened to Paul Mirecki?" December 7, 2005. Available online (January 2006) at: http://michelle-malkin.com/archives/2005_12.htm.

14. Michael Ruse, *Can a Darwinian Be a Christian?* (Cambridge: Cambridge University Press, 2001), ix, 28, 66–67, 82–83, 117, 217.

15. Kenneth R. Miller, *Finding Darwin's God: A Scientist's Search for Common Ground Between God and Evolution* (New York: HarperCollins, 2000), 167, 241.

16. Stephen Jay Gould, "Nonoverlapping Magisteria," *Natural History* 106 (March 1997): 16–22. Available online (June 2006) at: http://www.stephen-jaygould.org/library/gould_noma.html.

17. Alvin Plantinga, "Methodological Naturalism?—Part I," *Origins & Design* 18 (1997): 18–27. Available online (June 2006) at: http://www.arn.org/docs/odesign/od181/methnat181.htm.

18. Michael Foust, "Gallup Poll Latest to Show Americans Reject Secular Evolution," *Baptist Press* (October 19, 2005). Available online (June 2006) at: http://www.bpnews.net/bpnews.asp?ID=21891. "Poll: Creationism Trumps Evolution," CBS News, November 22, 2004. Available online (June 2006) at: http://www.cbsnews.com/stories/2004/11/22/opinion/polls/main657083.shtml. Ontario Consultants on Religious Tolerance, "Religious Identification in the U.S.," (2001). Available online (June 2006) at: http://www.religioustoler-ance.org/chr_prac2.htm. Eugenie C. Scott, "Dealing with Anti-Evolutionism,"

University of California–Berkeley Museum of Paleontology website. Available online (June 2006) at: http://www.ucmp.berkeley.edu/fosrec/Scott2.html.

19. Thomas J. Oord and Eric Stark, "A Conversation with Eugenie Scott," *Science and Theology News*, April 1, 2002. Available online (June 2006) at: http://www.stnews.org/Commentary-1835.htm.

20. Michael Zimmerman, "Welcome to the Clergy Letter Project" (January 5, 2006). Available online (June 2006) at: http://www.uwosh.edu/colleges/cols/clergy_project.htm. "The Clergy Letter Project Present Evolution Sunday," Feburary 12, 2006. Available online (June 2006) at: http://www.uwosh.edu/colleges/cols/rel_evol_sun.htm. Neela Banerjee and Anne Berryman, "At Churches Nationwide, Good Words for Evolution," *New York Times*, February 13, 2006. Available online (registration required; April 2006) at: http://www.nytimes.com/2006/02/13/national/13evolution.html?ex=1140498000&en=e4953893104c14d8&ei=5070&emc=eta1. Aaron Ricadela, "Darwin Takes the Pulpit on Evolution Sunday," Science & Theology News, May 4, 2006. Available online (June 2006) at: http://www.stnews.org/news-2810.htm.

21. Rob Crowther, "Is the Pope Catholic?" *Evolution News and Views*, May 31, 2006. Available online (June 2006) at: http://www.evolutionnews.org/2006/05/is_the_pope_catholic.html.

22. Pope Pius XII, *Humani Generis*, August 12, 1950. Available online (June 2006) at: http://academic.regis.edu/mghedott/humanigeneris.htm.

23. Pope John Paul II, "Message to the Pontifical Academy of Sciences," October 22, 1996. Available online (June 2006) at: http://www.ewtn.com/library/PAPALDOC/JP961022.HTM.

24. International Theological Commission, "Communion and Stewardship: Human Persons Created in the Image of God" (July 2004). Available online (June 2006) at: http://academic.regis.edu/mghedott/communionsteward-ship.htm and http://www.vatican.va/roman_curia/congregations/cfaith/cti_documents/rc_con_cfaith_doc_20040723_communion-stewardship_en.html.

25. Pope Benedict XVI, "Inaugural Address," *Boston Catholic Journal,* April 22, 2005. Available online (June 2006) at: http://www.boston-catholic-journal.com/inaugural_address_of_Pope_Benedict_XVI.htm.

26. Christoph Cardinal Schönborn, "Finding Design in Nature," *New York Times*, July 7, 2005. Available online (June 2006) at http://stephanscom.at/edw/reden/0/articles/2005/07/08/a8795/.

27. George Coyne, "God's Chance Creation," *Tablet* (August 6, 2005). Available online (June 2006) at: http://www.thetablet.co.uk/cgi-bin/register.cgi/tablet-01063. Martin Hilbert, "Darwin's Divisions: The Pope, the Cardinal, the Jesuit & the Evolving Debate About Origins," *Touchstone,* June 2006. Available online (June 2006) at: http://www.touchstonemag.com/archives/article.php?id=19-05-028-f.

28. Constance Holden, "Vatican Astronomer Rebuts Cardinal's Attack on Darwinism," *Science* 309 (2005): 996–97.

29. University of California Museum of Paleontology, "Misconception: Evolution and religion are incompatible," *Understanding Evolution.* Available online (June 2006) at: http://evolution.berkeley.edu/evosite/misconceps/IVAandreligion.shtml. Award Information, Grant 0096613, National Science Foundation (April 2001). Available online (June 2006) at: http://learn.arc.nasa.gov/benchmark/docs/NSF/Teacher%20Enhancement.doc.

30. "Statements from Religious Organizations," National Center for Science Education. Available online (June 2006) at: http://www.ncseweb.org/resources/articles/5025_statements_from_religious_orga_12_19_2002.asp.

31. John G. West, "Evolving Double Standards: Establishing a State-Funded Church of Darwin," *National Review*, April 1, 2004. Available online (June 2006) at: http://www.nationalreview.com/comment/west200404010900.asp. Francis J. Beckwith, "Government-Sponsored Theology," *American Spectator*, April 7, 2004. Available online (June 2006) at: http://www.spectator.org/dsp_article.asp?art_id=6395.

32. "Feds Fund Religious Promotion of Evolution," *WorldNetDaily*, October 13, 2005. Available online (June 2006) at: http://worldnetdaily.com/news/article.asp?ARTICLE_ID=46807. Casey Luskin, "Dismissal of Lawsuit against Evolution Website Implies Internet is an Establishment-Clause-Free-Zone," *Evolution News & Views*, March 29, 2006. Available online (June 2006) at: http://www.evolutionnews.org/2006/03/its_not_over_federal_judge_dis.html.

Chapter Sixteen:
American Lysenkoism

1. William A. Dembski, *The Design Revolution: Answering the Toughest Questions About Intelligent Design* (Downer's Grove, IL: InterVarsity Press, 2004), 304–05.

2. Wesley R. Elsberry and Mark Perakh, "How Intelligent Design Advocates Turn the Sordid Lessons from Soviet and Nazi History Upside Down," April 20, 2004. Available online (January 2006) at: http://www.talkreason.org/articles/eandp.cfm.

3. Robert T. Pennock, "DNA by Design?" *Debating Design: From Darwin to DNA*, ed.William A. Dembski and Michael Ruse, (Cambridge: Cambridge University Press, 2004), 130–48. Douglas J. Futuyma, *Evolution* (Sunderland, MA: Sinauer Associates, 2005), 537.

4. Christopher Scott, "Stand By Science," *Edutopia*, June 1, 2005. Available online (January 2006) at: http://www.edutopia.org/magazine/ed1article.php?id=Art_1314&issue=jun_05.

5. Nils Roll-Hansen, *The Lysenko Effect: The Politics of Science* (Amherst, NY: Humanity Books, 2005), 86–88.

6. Darwin, *On the Origin of Species*, Sixth Edition, Chapter V. See also Chapters I and VI.

7. B. C. A. Windle, "Mendel, Mendelism," *Catholic Encyclopedia,* Volume X (Robert Appleton Company, 1911). Available online (January 2006) at: http://www.newadvent.org/cathen/10180b.htm. Peter J. Bowler, *Evolution: The History of an Idea*, Revised Edition (Berkeley, CA: University of California Press, 1989), 171, 190, 210.

8. L. C. Dunn, *A Short History of Genetics* (Ames, IA: Iowa State University Press, 1991), Chapter 1. Jan Sapp, *Beyond the Gene: Cytoplasmic Inheritance and the Struggle for Authority in Genetics.* (New York: Oxford University Press, 1987), Chapters 2–4. Loren R. Graham, *What Have We Learned about Science and Technology from the Russian Experience?* (Stanford, CA: Stanford University Press, 1998), 19–23.

9. Roll-Hansen, *The Lysenko* Effect, 12–16. Graham, *What Have We Learned about Science and Technology from the Russian Experience?*, xi–xii, 53, 132–33.

10. Roll-Hansen, *The Lysenko Effect,* Chapter 3. David Joravsky, *The Lysenko Affair* (Cambridge, MA: Harvard University Press, 1970), Chapter 3. Valery N. Soyfer, *Lysenko and the Tragedy of Soviet Science* (New Brunswick, NJ: Rutgers University Press, 1994), Chapters 1, 2, and Conclusion.

11. Roll-Hansen, *The Lysenko Effect*, 86–89. Soyfer, *Lysenko and the Tragedy of Soviet Science*, 63. Joravsky, *The Lysenko Affair,* 208, 238–39.

Zhores Medvedev, *The Rise and Fall of T. D. Lysenko* (New York: Columbia University Press, 1969), Chapter 3.

12. Roll-Hansen, *The Lysenko Effect,* 218–20. Medvedev, *The Rise and Fall of T. D. Lysenko,* 46–49.

13. Medvedev, *The Rise and Fall of T. D. Lysenko,* Chapter 11. Graham, *What Have We Learned about Science and Technology from the Russian Experience?* Chapter 1 and Conclusions. Roll-Hansen, *The Lysenko Effect,* Chapter 10.

14. Elsberry and Perakh, "How Intelligent Design Advocates Turn the Sordid Lessons from Soviet and Nazi History Upside Down."

15. National Science Teachers Association, "Survey Indicates Science Teachers Feel Pressure to Teach Nonscientific Alternatives to Evolution," March 24, 2005. Available online (January 2006) at: http://science. nsta.org/nstaexpress/nstaexpress_2005_03_28_pressrelease.htm.

16. Michelle Malkin, "What Happened to Paul Mirecki?" December 7, 2005. Available online (January 2006) at: http://michellemalkin.com/ archives/2005_12.htm.

17. The book is Francis J. Beckwith, *Law, Darwinism & Public Education: The Establishment Clause and the Challenge of Intelligent Design* (Lanham, MD: Rowman & Littlefield, 2003). Lawrence VanDyke, "Book Note: Not Your Daddy's Fundamentalism: Intelligent Design in the Classroom," *Harvard Law Review* 117 (January, 2004), 964. Brian R. Leiter, "Harvard Law Review Embarrasses Itself," March 10, 2004, available online (January 2006) at: http://leiterreports.typepad.com/blog/2004/03/ harvard_law_rev.html. Hunter Baker, "The Professor's Paroxysm," National Review Online, March 15, 2004. Available online (January 2006) at: http://www.nationalreview.com/comment/baker200403150909.asp.

18. Catherine Candisky, "Evolution Debate Re-emerges: Doctoral Student's Work was Possibly Unethical, OSU Professors Charge," *Columbus Dispatch,* June 9, 2007. Annie Hall, "Intelligent Design Debate Continues on National Level," *Lantern,* January 3, 2006, 1. Available online (January 2006) at: http://www.thelantern.com/media/paper333/ news/2006/01/03/Campus/Intelligent.Design.Debate.Continues.On.Nationa l.Level1270092.shtml?norewrite&sourcedomain=www.thelantern.com. Letter by Professor Robert DiSilvestro (a member of Leonard's dissertation committee) in *The Scientist,* August 29, 2005, 8. McKee was quoted by

Robert Crowther on the blog "Evolution News & Views," January 10, 2006. Available online (January 2006) at: http://www.evolutionnews.org/2006/ 01/in_ohio_darwinist_admits_plan.html.

19. Myers's blogs are available online (January 2006) at: http://www.pandasthumb.org/archives/2005/06/a_new_recruit.html#c35130 and http://pharyngula.org/index/weblog/comments/perspective/P25/.

20. Jim Brown and Ed Vitagliano, "Professor Dumped Over Evolution Beliefs," Agape Press, March 11, 2003. Available online (January 2006) at: http://headlines.agapepress.org/archive/3/112003a.asp.

21. Barbara Bradley Haggerty, "Profile: Intelligent Design and Academic Freedom," transcript of *All Things Considered* (NPR), November 10, 2005. Available online (January 2006) at: http://www.discovery.org/scripts/ viewDB/index.php?command=view&program=CSC%20-%20Views%20 and%20News&id=3083. Geoff Brumfiel, "Cast Out of Class," *Nature* 434, April 28, 2005, 1062–65. Available online (January 2006) at: http://www.nature.com/nature/journal/v434/n7037/box/4341062a_bx1.html. Shankar Vedantam, "Eden and Evolution," *Washington Post*, February 5, 2006. Available online (April 2006) at: http://www.washingtonpost. com/wp-dyn/content/article/2006/02/03/AR2006020300822.html.

22. Judge Jones's decision is available online (January 2006) at http://www.pamd.uscourts.gov/kitzmiller/kitzmiller_342.pdf. For a detailed analysis of the decision, see John G. West, "Dover in Review," Discovery Institute, January 6, 2006. Available online (January 2006) at: http://www.discovery.org/scripts/viewDB/index.php?command=view&id= 3135&program=CSC%20-%20Science%20and%20Education%20Policy%20-%20News%20and%20Articles.

Chapter Seventeen:
Scientific Revolution

1. David Donnenfield and David Howell, "The Birth of Plate Tectonics Theory," *U.S. Geological Survey*, April 26, 2006. Available online (June 2006) at: http://www.usgs.gov/125/articles/plate_tectonics.html.

2. Thomas S. Kuhn, *The Structure of Scientific Revolutions,* Revised Edition (Chicago: The University of Chicago Press, 1970), 5–6, 10, 34, 175. Imre Lakatos and Alan Musgrave, *Criticism and the Growth of Knowledge* (Cambridge: Cambridge University Press, 1970).

3. Kuhn, 24, 92, 145.

4. Kuhn, 77.

5. Kuhn, 94.

6. Kuhn, 171–73, 205–06.

7. Tim McGrew, "Scientific Progress, Relativism, and Self-Refutation," *Electronic Journal of Analytic Philosophy* 2 (1994). Available online (June 2006) at: http://ejap.louisiana.edu/EJAP/1994.may/mcgrew.html. James Franklin, "Thomas Kuhn's irrationalism," *New Criterion* 18 (June 2000). Available online (June 2006) at: http://www.newcriterion.com/archive/18/jun00/kuhn.htm.

8. Kuhn, 91, 103–05, 148, 163.

9. Ibid., 93–96.

10. Ibid., 157–58.

11. Ibid., 18–19, 90, 144, 151, 166–67.

12. Scott Todd, "A view from Kansas on that evolution debate," *Nature* 401 (1999): 423.

13. Orson Scott Card, "Creation and Evolution in the Schools," *Ornery American* (January 8, 2006). Available online (June 2006) at: http://www.ornery.org/essays/warwatch/2006-01-08-1.html.

14. William Rusher, "What are the scientists really afraid of?" *Decatur Daily Democrat* (January 10, 2006). Available online (June 2006) at: http://www.decaturdailydemocrat.com/articles/2006/01/10/news/opinion/editorial03.txt.

15. Douglas Kern, "Why Intelligent Design Is Going to Win," *TCS Daily*, October 7, 2005. Available online (June 2006) at: http://www.tcsdaily.com/article.aspx?id=100705C.

16. Mark Perakh, "The dream world of William Dembski's creationism," *Skeptic Magazine* 11, March 22, 2005, 54–65. Richard Dawkins and Jerry Coyne, "One side can be wrong," *Guardian*, September 1, 2005. Available online (June 2006) at: http://www.guardian.co.uk/life/feature/story/0,13026,1559743,00.html.

17. Jodi Wilgoren, "Politicized Scholars Put Evolution on the Defensive," *New York Times*, August 21, 2005. Available online (June 2006) at: http://www.nytimes.com/2005/08/21/national/21evolve.html?ex=1150171200&en=8215596daed2bf03&ei=5070.

18. Molecular & Microbiology Faculty, George Mason University. Available online (June 2006) at: http://www.gmu.edu/departments/mmb/fac-

ulty.html. "Table 4: Average Salary and Average Compensation Levels, 2005–2006," *Annual Report on the Economic Status of the Profession, 2005–06*, American Association of University Professors, March–April, 2006. Available online (June 2006) at: http://www.aaup.org/surveys/06z/alltabs.pdf.

19. Award Abstract #0229294, "Explore Evolution," National Science Foundation, November 29, 2002. Available online (June 2006) at: http://www.nsf.gov/awardsearch/showAward.do?AwardNumber=0229294. "NSF Budget Is at $5.58 Billion for FY 2006," National Science Foundation January 12, 2006. Available online (June 2006) at: http://www.nsf.gov/about/congress/109/highlights/cu06_0112.jsp. Jeffrey Mervis, "Senate Panel Chair Asks Why NSF Funds Social Sciences," *Science* 312 (2006), 829.

20. "Summary of the FY 2006 President's Budget," *National Institutes of Health*, February 7, 2005. Available online (June 2006) at: http://www.nih.gov/news/budget/FY2006presbudget.pdf. Warren E. Johnson, Eduardo Eizirik, Jill Pecon-Slattery, William J. Murphy, Agostino Antunes, Emma Teeling, and Stephen J. O'Brien, "The Late Miocene Radiation of Modern Felidae: A Genetic Assessment," *Science* 311 (2006), 73-77.

21. William A. Dembski, "Becoming a Disciplined Science: Prospects, Pitfalls, and Reality Check for ID," Keynote address delivered at RAPID Conference (Research and Progress in Intelligent Design), Biola University, La Mirada, CA, October 25, 2002. Available online (June 2006) at: http://www.designinference.com/documents/2002.10.27.Disciplined_Science.htm. Wolf-Ekkehard Lönnig, "Dynamic genomes, morphological stasis and the origin of irreducible complexity," *Dynamical Genetics* (Kerala, India: Research Signposts, 2004), 101–19. "Peer-Reviewed & Peer-Edited Scientific Publications Supporting the Theory of Intelligent Design (Annotated)," Discovery Institute, February 22, 2006. Available online (June 2006) at: http://www.discovery.org/scripts/viewDB/index.php?command=view&id=2640&program=CSC%20-%20Scientific%20Research%20and%20Scholarship%20-%20Science.

22. Michael J. Behe and David W. Snoke, "Simulating evolution by gene duplication of protein features that require multiple amino acid residues," *Protein Science* 13 (2004), 2651–2664.

23. Douglas D. Axe, "Extreme Functional Sensitivity to Conservative Amino Acid Changes on Enzyme Exteriors," *Journal of Molecular Biology*

301 (2000), 585–95. Douglas D. Axe, "Estimating the Prevalence of Protein Sequences Adopting Functional Enzyme Folds," *Journal of Molecular Biology* 341 (2004), 1295–1315.

24. Forrest M. Mims III, "Avian Influenza and UV-B Blocked by Biomass Smoke," *Environmental Health Perspectives* 113, December 2005. Available online (June 2006) at: http://www.ehponline.org/realfiles/docs/2005/113-12/correspondence.html.

25. Jonathan Wells, "Do Centrioles Generate a Polar Ejection Force?" *Rivista di Biologia—Biology Forum* 98 (2005), 71–96. Available online (June 2006) at: http://www.discovery.org/scripts/viewDB/index.php?command=view&id=2680. Jonathan Wells, "Using Intelligent Design Theory to Guide Scientific Research," *International Society for Complexity, Information, and Design*, May 13, 2004. Available online (June 2006) at: http://www.iscid.org/boards/ubb-get_topic-f-10-t-000081.html.

26. Hervé Morin, "Le darwinisme est parfois contesté dans les salles de classe françaises," *Le Monde*, April 27, 2005. Available online (June 2006) at: http://www.lemonde.fr/cgibin/ACHATS/acheter.cgi?offre=ARCHIVES&type_item=ART_ARCH_30J&objet_id=898462. "Darwin and Design," Prague, Czech Republic, October 22, 2005. Program available online (June 2006) at: http://www.darwinanddesign.org/." Design of Creation Society," Japan. Available online (June 2006) at: http://www.dcsociety.org/. "Núcleo Brasileiro de Design Inteligente," Brazil. Website at: http://www.nbdi.org.br. "Desafiando a Nomenklatura Científica (Defying the Scientific Nomenklatura)," Brazil. Available online (June 2006) at: http://pos-darwinista.blogspot.com/.

27. IDEA Club, Cornell University. Website available online (June 2006) at: http://www.rso.cornell.edu/idea/whatwebelieve.html. Lisa Anderson, "Students join debate on intelligent design: Campus clubs set up to defend concept," *Chicago Tribune*, November 25, 2005. Kerry Frisinger, "Evolution debate PRI director lauds Pa. intelligent design ruling," *Ithaca Journal*, December 23, 2005. Available online (June 2006) at: http://www.theithacajournal.com/apps/pbcs.dll/article?AID=/20051223/NEWS01/512230308/1002.

INDEX

Minnesota, University of, 191

Minnich, Scott, 115–17, 128

miracles, 133–34

Mirecki, Paul, 173, 174, 187

missing links: Darwinism and, 13; evolution and, 21, 22

Mississippi University for Women, 190, 191

molecular phylogeny, 37–39; biology textbooks and, 40, 43; Darwinism and, 37–47, 38–39; descent with modification and, 47; evolutionary tree and, 37, 38; molecular dating and, 46–47; tree of life and, 43, 44–46; universal common ancestry and, 44–46; whale evolution and, 40–41; whippo hypothesis and, 40–41

Monkey Shakespeare Simulator, 92–93

Monod, Jacques, 5, 95

Morris, Simon Conway, 17

Muizon, Christian de, 41

Müntzing, Arne, 51

mutations: DNA and, 95; embryology and, 33; natural selection and, 2

Myers, Paul Z., 186, 190, 191

The Mystery of Life's Origin (Thaxton, Bradley, and Olsen), 9, 98

N

National Cancer Institute, 202

National Center for Biotechnology Information, 43

National Center for Science Education (NCSE), 28–29, 66, 105, 144

National Institutes of Health, 202

National Museum of Natural History (NMNH), 103, 124–25, 126

National Review, 188

National Science Foundation (NSF), 178, 202

National Science Teachers Association, 187

Natural History, 32, 144

natural law, 135; descent with modification and, 135; intelligent design and, 9; science and, 132

natural selection: biological information, origin of and, 102; Darwinism and, 3–6; descent with modification and, 3; evolution and, 14; intelligent design and, 8; irreducible complexity and, 109–10; mutations and, 2; neo-Darwinism and, 146–47; variations and, 6. *See also* Darwinism

natural theology, intelligent design and, 8

naturalism, methodological, 133–35

Nature, 22, 35, 57, 102, 103, 144, 151

Nature, Design, and Science: The Status of Design in Natural Science (Ratzsch), 134

"The Nature of Nature" conference (2000), 89–91

Nature's Destiny (Denton), 120

NCSE. *See* National Center for Science Education

Nelson, Gareth, 18, 21

Nelson, Paul A., 68

neo-Darwinism: divine providence and, 177; embryology and, 33; intelligent design and, 186; Lysenkoism and, 182; Mendelism and, 73; natural selection and, 146–47. *See also* Darwinism

Nernst, Walther, 137, 138

New Testament, 170

Other Politically Incorrect Guides™

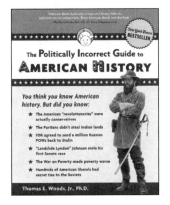

0-89526-047-6, $19.95, paperback

0-89526-013-1, $19.95, paperback

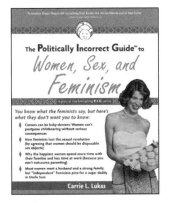

1-59698-003-6, $19.95, paperback

0-89526-031-X, $19.95, paperback

Regnery Publishing created the bestselling Politically Incorrect Guide™ (P.I.G.) series to tackle a variety of hot topics in our society—issues that have been hijacked by politically correct historians, academia, and media. Inside every P.I.G. you'll find politically correct myths busted with an abundance of cold, hard facts.

Look for these upcoming P.I.G.s, including:

The Politically Incorrect Guide™ to English and American Literature
The Politically Incorrect Guide™ to the South (and Why It Should Rise Again)
The Politically Incorrect Guide™ to Global Warming (and Environmentalism)